High-Yield Biochemistry

High-Yield Biochemistry

R. Bruce Wilcox, Ph.D
Professor of Biochemistry
Loma Linda University
Loma Linda, California

LIPPINCOTT WILLIAMS & WILKINS
A **Wolters Kluwer** Company
Philadelphia · Baltimore · New York · London
Buenos Aires · Hong Kong · Sydney · Tokyo

Editor: Elizabeth A. Nieginski
Managing Editor: Marette D. Magargle-Smith
Marketing Manager: Jennifer Conrad
Development Editor: Martha Cushman

351 West Camden Street
Baltimore, Maryland 21201-2436 USA

Rose Tree Corporate Center
1400 North Providence Road
Building II, Suite 5025
Media, Pennsylvania 19063-2043 USA

Printed in the United States of America

Library of Congress Cataloging-in-Publication Data
Wilcox, R. Bruce.
    High-yield Biochemistry / R. Bruce Wilcox.
        p.   cm. — (High-yield series)
    ISBN 0-683-30459-3
    1. Biochemistry.   I. Title.   II. Series.
QP514.2.W52   1998
572—dc21
                                                      98-38501
                                                            CIP

The publishers have made every effort to trace the copyright holders for borrowed material. If they have inadvertently overlooked any, they will be pleased to make the necessary arrangements at the first opportunity.

To purchase additional copies of this book, call our customer service department at **(800) 638-0672** or fax orders to **(800) 447-8438.** For other book services, including chapter reprints and large quantity sales, ask for the Special Sales department.

Canadian customers should call **(800) 665-1148,** or fax **(800) 665-0103.** For all other calls originating outside of the United States, please call **(410) 528-4223** or fax us at **(410) 528-8550.**

Visit Williams & Wilkins on the Internet: **http://www.wwilkins.com** or contact our customer service department at **custserv@wwilkins.com**. Williams & Wilkins customer service representatives are available from 8:30 am to 6:00 pm, EST, Monday through Friday, for telephone access.

                                                      99 00 01 02
                                                 2 3 4 5 6 7 8 9 10

Dedication

This book is dedicated to my father, H. Bruce Wilcox, for endowing me with a passionate love for teaching, and to the freshman medical and dental students at Loma Linda University who for over 30 years have paid tuition at confiscatory rates so that I have never had to go to work.

Contents

Preface

High-Yield Biochemistry is based on a series of notes prepared in response to repeated and impassioned requests by my students for a "complete and concise" review of biochemistry. It is designed for rapid review during the last days and hours before the United States Medical Licensing Examination (USMLE), Step 1. While this book provides the essential information needed for a quick review, always remember that you cannot review what you never knew.

Acknowledgments

I would like to thank Dr. John Sands for his invaluable help in reviewing and editing Chapter 11, "Biotechnology." Many thanks are due to the staff at Lippincott Williams & Wilkins, especially Associate Development Editor, Lisa Bolger, for her many encouraging e-mail messages through what has truly been a long, hot summer. I am also indebted to Dr. J. Paul Stauffer at Pacific Union College for instruction in the felicitous use of English, to the late P.G. Wodehouse for continuing and enriching that instruction, and to General U.S. Grant for providing an example of laconic communication.

1

Acid–Base Relationships

I. ACIDIC DISSOCIATION

A. An **acid dissociates** in water to yield a **hydrogen ion (H^+)** and its **conjugate base.**

$$\begin{array}{cc} \text{Acid} & \text{Conjugate base} \\ \text{(acetic acid)} & \text{(acetate)} \\ CH_3COOH \underset{H_2O}{\overset{}{\rightleftharpoons}} & H^+ + CH_3COO^- \end{array}$$

B. A **base combines** with H^+ in water to form its **conjugate acid.**

$$\begin{array}{cc} \text{Ammonia} & \text{Ammonium ion} \\ \text{(base)} & \text{(conjugate acid)} \\ NH_3 + H^+ \underset{H_2O}{\overset{}{\rightleftharpoons}} & NH_4^+ \end{array}$$

C. In the more general expression of **acidic dissociation, HA is the acid** (proton donor) and **A^- is the conjugate base** (proton acceptor).

$$HA \underset{k_{-1}}{\overset{k_1}{\rightleftharpoons}} H^+ + A^-$$

where $k_1[HA]$ = the forward rate and $k_{-1}[H^+][A^-]$ = the reverse rate.

II. MEASURES OF ACIDITY

A. pK_a

    1. When the forward and reverse rates are equal (see the equation for acidic dissociation), the **acidic dissociation constant, K_a,** is defined by:

$$\frac{k_1}{k_{-1}} = \frac{[H^+][A^-]}{[HA]} = K_a$$

2. pK_a expresses the **strength** of an acid. *(or a high K_a)*

a. By definition, pK_a equals $-\log[K_a]$.

b. A **strong acid** has a **pK_a of 2 or less,** which indicates that H^+ binds loosely to the conjugate base. Examples of strong acids include hydrochloric acid (HCl) and sulfuric acid (H_2SO_4). *(low K_a)*

c. A **weak acid** has a **pK_a of 10 or more,** which indicates that H^+ binds tightly to the conjugate base. Examples of weak acids include acetic acid and citric acid.

B. pH

1. When the equation defining K_a is further rearranged and expressed in logarithmic form, it becomes the **Henderson-Hasselbalch equation:**

$$pH = pK_a + \log\frac{[A^-]}{[HA]}$$

2. pH is a measure of the **acidity** of a solution.

a. By definition, **pH equals $-\log[H^+]$.**

b. A **neutral solution** has a $[H^+]$ of 10^{-7}, which means that it has a **pH of 7.**

c. An **acidic solution** has a $[H^+]$ **greater than** 10^{-7}, which means that it has a **pH of less than 7.**

d. An **alkaline solution** has a $[H^+]$ **less than** 10^{-7}, which means that it has a **pH of greater than 7.**

III. BUFFERS

A. A **buffer** is a solution that contains a mixture of a weak acid and its conjugate base. It resists changes in $[H^+]$ on addition of acid or alkali.

B. The **buffering capacity** of a solution is determined by the acid–base **concentration** and the **pK_a** of the weak acid.

1. The **maximum buffering effect** occurs when the concentration of the weak acid is equal to that of its conjugate base.

2. When the buffer effect is at its maximum, the **pH of the solution equals the pK_a of the acid.**

C. The buffering effect is readily apparent on the **titration curve** for a weak acid such as $H_2PO_4^-$ (Figure 1-1).

1. The **shape** of the titration curve is the same for all weak acids.

2. At the **midpoint** of the curve, **the pH equals the pK_a.** $\left([A^-] = [HA] \right)$

3. The **buffering region** extends **one pH unit** above and below the pK_a.

IV. ACID–BASE BALANCE

A. Because pH strongly affects the stability of proteins and the catalytic activity of enzymes, **biological systems usually function best at a $[H^+]$ of 10^{-7} M or a near-neutral pH (approximately 7).** Under normal conditions, blood pH is 7.4 (range, 7.37–7.42).

B. The acid–base pair **monohydrogen phosphate ($H_2PO_4^-$)–dihydrogen phosphate (HPO_4^{2-})** is an **effective buffer pair in physiologic fluids** at normal pH (see Figure 1-1).

So you want to choose on acid 1) a pK_a close to the pH you 2) want to maintain

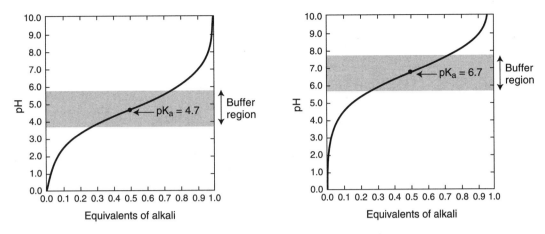

Figure 1-1. Titration curves for acetic acid (CH_3COOH) (*left*) and phosphoric acid ($H_2PO_4^-$) (*right*). $H_2PO_4^-$ is the more effective buffer at physiologic pH.

$$H_2PO_4^- \rightleftharpoons HPO_4^{2-} + H^+$$

$$pH = 6.7 + \log \frac{[HPO_4^{2-}]}{[H_2PO_4^-]}$$

C. The **carbon dioxide (CO_2)-carbonic acid (H_2CO_3)-bicarbonate (HCO_3^-) system** is the principal buffer in plasma and extracellular fluid (ECF).

$$CO_2 + H_2O \xrightleftharpoons{\text{Carbonic anhydrase}} H_2CO_3 \rightleftharpoons H^+ + HCO_3^-$$

1. CO_2 from tissue oxidation reactions dissolves in the blood plasma and ECF.

2. CO_2 combines with H_2O to yield H_2CO_3. This reaction is catalyzed in red blood cells by **carbonic anhydrase**.

3. H_2CO_3 dissociates to yield H^+ and its conjugate base, HCO_3^-.

4. In this system, CO_2 is behaving like an acid, so the Henderson-Hasselbalch equation can be written:

$$pH = 6.1 + \log \frac{[HCO_3^-]}{(0.0301) \, P_{CO_2}}$$

where $[HCO_3^-]$ is in mM and P_{CO_2} is in mm Hg.

D. The **CO_2-H_2CO_3-HCO_3^- buffer system** is effective around the physiologic pH of 7.4, even though the pK_a is only 6.1, for four reasons:

1. The supply of CO_2 from oxidative metabolism is unlimited, so the effective concentration of CO_2 is very high.

2. Equilibration of CO_2 with H_2CO_3 (catalyzed by carbonic anhydrase) is very rapid.

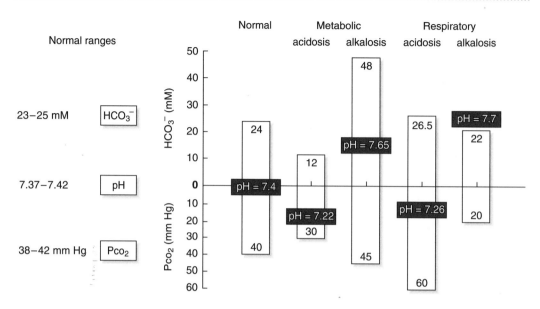

Figure 1-2. Bar chart that demonstrates prototypical acid–base states of extracellular fluid (ECF). HCO_3^- is plotted up from zero, and P_{CO_2} is plotted down from zero.

3. The variation in CO_2 removal by the lungs allows for rapid changes in the concentration of the H_2CO_3.

4. The kidney can retain or excrete HCO_3^-, thus changing the concentration of the conjugate base.

V. ACID–BASE DISORDERS

A. **Acidosis** occurs when the pH of the blood and ECF falls below 7.35 ($[H^+] > 44.7$ nmol/L). This condition results in **central nervous system depression**, and when severe, it can lead to coma and death.

   1. In **respiratory acidosis**, the **partial pressure of CO_2 (P_{CO_2}) increases** as a result of **hypoventilation** (Figure 1-2).

   2. In **metabolic acidosis**, the **$[HCO_3^-]$ decreases** as a consequence of the addition of an acid stronger than H_2CO_3 to the ECF.

B. **Alkalosis** occurs when the pH of the blood and ECF rises above 7.45 ($[H^+] < 35.5$ nmol/L). This condition leads to **neuromuscular hyperexcitability,** and when severe, it can result in tetany.

   1. In **respiratory alkalosis**, the P_{CO_2} **decreases** as a consequence of **hyperventilation**.

   2. In **metabolic alkalosis**, the **$[HCO_3^-]$ increases** as a consequence of excess acid loss (e.g., vomiting) or addition of a base (e.g., oral antacid preparation).

VI. CLINICAL RELEVANCE: DIABETIC KETOACIDOSIS

A. Uncontrolled **insulin-dependent diabetes mellitus (type I diabetes)** involves **decreased glucose utilization,** with hyperglycemia, and increased fatty acid oxidation.

B. Pathogenesis of ketoacidosis

  1. **Increased fatty acid oxidation** leads to excessive production of acetoacetic and 3-hydroxybutyric acids and acetone, which are known as **ketone bodies.**

  2. Acetoacetic and 3-hydroxybutyric **acids dissociate** at body pH and release H⁺, leading to a **metabolic acidosis.**

C. The combination of high blood levels of the ketone bodies and a metabolic acidosis is called **ketoacidosis.**

D. The **clinical picture** involves **dehydration, lethargy, and vomiting,** followed by drowsiness and coma.

E. **Therapy** consists of correction of the hyperglycemia, dehydration, and acidosis.

  1. **Insulin** is administered to correct the hyperglycemia.

  2. **Fluids** in the form of physiologic saline are administered to treat the dehydration.

  3. In severe cases, intravenous **sodium bicarbonate** ($Na^+HCO_3^-$) may be administered to correct the acidosis.

2

Amino Acids and Proteins

I. FUNCTIONS OF PROTEINS

   A. Specific binding to other molecules

   B. Catalysis

   C. Structural support

   D. Coordinated motion

II. PROTEINS AS POLYPEPTIDES

   A. Proteins are polypeptides, which are **linear polymers of amino acids** linked together by **peptide bonds** (Figure 2-1).

   1. Proteins are synthesized from **20 different amino acids.**

   2. Some of the amino acids are modified after incorporation into proteins (e.g., by hydroxylation, carboxylation, or phosphorylation). This is called **post-translational modification.**

   B. The amino acids are called **α-amino acids** because they have an **amino (–NH$_2$)** group, a **carboxyl (–COOH)** group, and some other "R-group" attached to the α-carbon (see Figure 2-1).

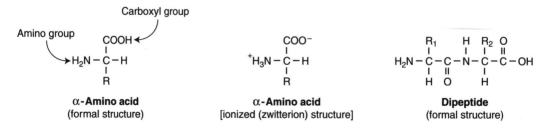

Figure 2-1. Structure of an α-amino acid and a dipeptide.

1. **Aliphatic R-groups** that are all **nonpolar (uncharged, hydrophobic)** [Figure 2-2] are characteristic of alanine, valine, leucine, isoleucine, and proline, which is an imino acid (a secondary amine). Glycine does not have a side chain.

2. **Aromatic R-groups** that are all **nonpolar** are components of phenylalanine, tyrosine, and tryptophan (see Figure 2-2).

3. **Hydroxyl-containing R-groups** that are **mildly polar (uncharged, hydrophilic)** are part of serine and threonine (see Figure 2-2).

4. **Sulfur-containing R-groups** are characteristic of cysteine (a good reducing agent) and methionine (see Figure 2-2).

5. **Carbonyl-containing R-groups** include the **carboxylates** aspartic acid and glutamic acid and their **amides** asparagine and glutamine. The carboxylates are **negatively charged and polar,** and their amides are **uncharged and mildly polar** (see Figure 2-2).

6. **Basic R-groups,** which are **positively charged** and **polar (hydrophilic),** are characteristic of lysine, arginine, and histidine (see Figure 2-2).

C. Each protein has a characteristic shape, or **conformation.**

   1. **The function of a protein is a consequence of its conformation.** The conformation of a functional protein is also called its **native structure.**

   2. The **amino acid sequence** of a protein determines its conformation.

   a. The **rigid, planar nature of peptide bonds** dictates the conformation that a protein can assume.

   b. The **nature and arrangement of the R-groups** further determines the conformation.

III. PROTEIN STRUCTURE. Four levels of hierarchy in protein conformation can be described.

   A. **Primary structure** refers to the order of the amino acids in the peptide chain (Figure 2-3).

   1. The **free α-amino group,** which is written to the left, is called the **amino-terminal** or **N-terminal** end.

   2. The **free α-carboxyl group,** which is written to the right, is called the **carboxyl-terminal** or **C-terminal** end.

   B. **Secondary structure** is the arrangement of hydrogen bonds between the peptide nitrogens and the peptide carbonyl oxygens of different amino acid residues (Figure 2-4; see Figure 2-3).

   1. In **helical coils,** the hydrogen-bonded nitrogens and oxygens are on nearby amino acid residues (see Figure 2-3).

   a. The most common helical coil is a right-handed **α-helix.**

   b. **α-keratin** from hair and nails is an α-helical protein.

   c. **Myoglobin** has several α-helical regions.

   d. Proline, glycine, and asparagine are seldom found in α-helices—they are "helix breakers."

   2. In **β-sheets (pleated sheets),** the hydrogen bonds occur between residues on neighboring peptide chains (see Figure 2-3).

   a. The hydrogen bonds may be on different chains or distant regions of the same chain.

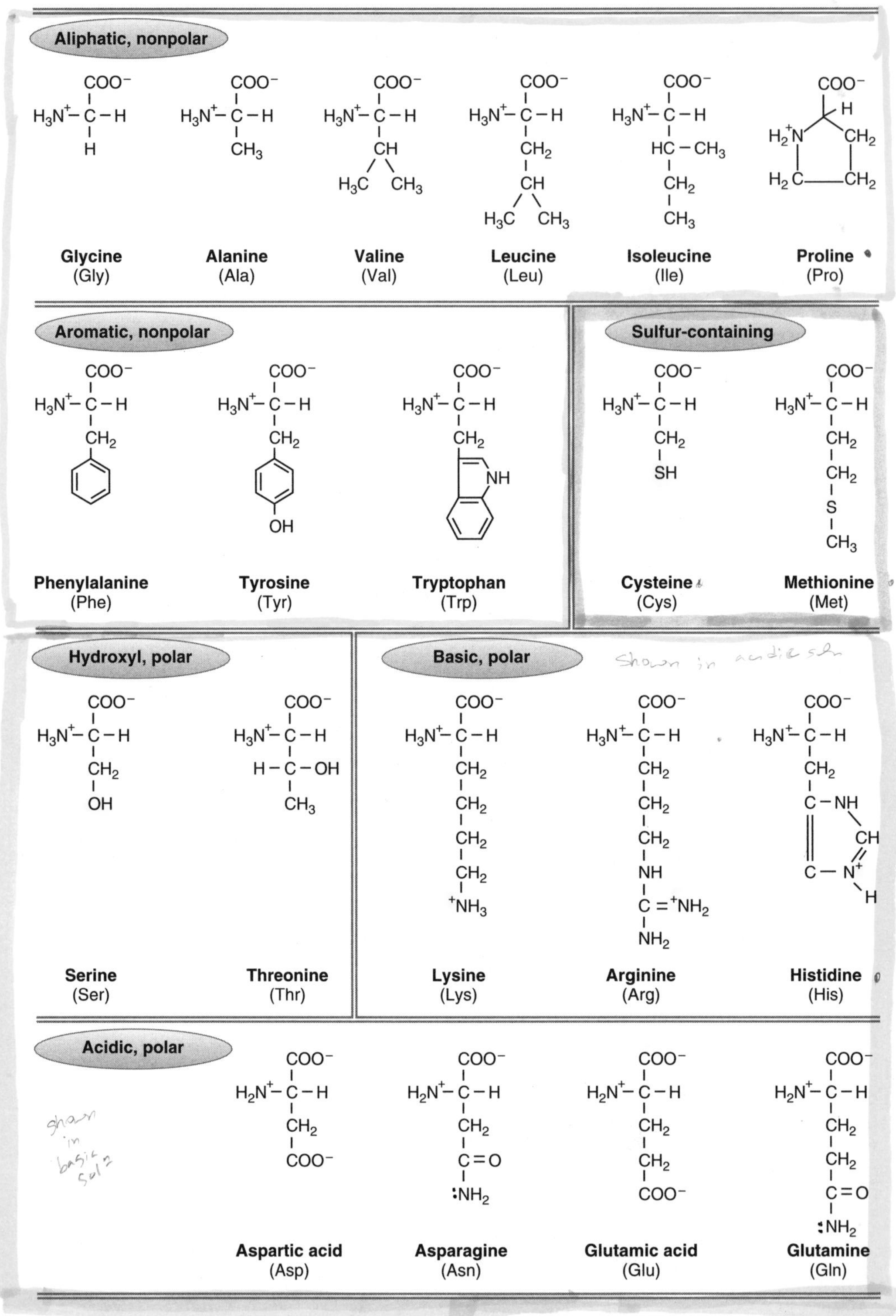

Figure 2-2. The 20 amino acids found in proteins, grouped by the properties of their R-groups.

Primary structure (always written with the free amino group to the left):

$$H_3N^+-\underset{\underset{R_1}{|}}{C}-\underset{\underset{}{\overset{\overset{O}{\|}}{C}}}-\underset{\underset{H}{|}}{N}-\underset{\underset{R_2}{|}}{C}-\underset{\underset{}{\overset{\overset{O}{\|}}{C}}}-\!\!\!-\!\!\!-\underset{\underset{H}{|}}{N}-\underset{\underset{R_n}{|}}{C}-COO^-$$

Secondary structures

α-Helix
(intramolecular
hydrogen bonds within
one polypeptide chain)

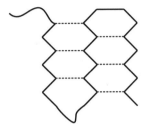

β-Pleated sheet
(intramolecular
hydrogen bonds within
one polypeptide chain)

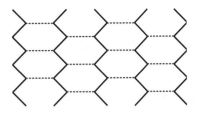

β-Pleated sheet
(intermolecular
hydrogen bonds between
different polypeptide chains)

Tertiary structure

Quaternary structure

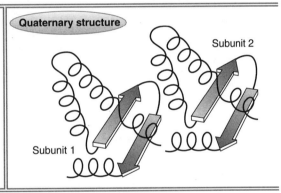

Subunit 2

Subunit 1

Figure 2-3. The four levels of protein structure.

$$\cdots\cdots\underset{\underset{H}{|}}{N}-CH_2-\underset{\underset{\underset{O}{\|}}{}}{\overset{\overset{R_1}{|}}{C}}-\underset{\underset{H}{|}}{N}-\underset{\underset{R_2}{|}}{CH_2}\cdots\cdots\rightarrow$$

← Hydrogen bond

$$\cdots\cdots\underset{\underset{H}{|}}{N}-CH_2-\underset{\underset{\underset{O}{\|}}{}}{\overset{\overset{R_3}{|}}{C}}-\underset{\underset{H}{|}}{N}-\underset{\underset{R_4}{|}}{CH_2}\cdots\cdots\rightarrow$$

Figure 2-4. A hydrogen bond between a peptide carbonyl oxygen and amide nitrogen.

  b. The strands may run **parallel** or **antiparallel.**

  c. **Fibroin in silk** is a β-sheet protein.

 3. β-bends, which are sharp bends in the polypeptide chain, connect regions of an α-helix or β-sheet.

 4. **Left-handed helical strands** are wound into a supercoiled **triple helix** in **collagen.** The major structural protein in the body, collagen makes up 25% of all vertebrate protein.

  a. The primary structure of collagen includes long stretches of the **repeating sequence glycine-X-Y,** where X and Y are frequently proline or lysine. The high proportion of **proline** residues leads to formation of the **left-handed** helical strands.

  b. Only **glycine** has an R-group small enough to fit into the interior of the **right-handed triple helix.**

  c. Collagen also contains **hydroxyproline** and **hydroxylysine.** The hydroxyl groups are added to proline and lysine residues by post-translational modification.

C. **Tertiary structure** refers to the **three-dimensional arrangement** of a polypeptide chain that has assumed its secondary structure (see Figure 2-3). <u>Disulfide bonds between cysteine</u> ✳ <u>residues may stabilize tertiary structure.</u>

D. **Quaternary structure** is the arrangement of the subunits of a protein that has more than <u>one polypeptide chain</u> (see Figure 2-3).

IV. PROTEIN SOLUBILITY AND R-GROUPS

A. **Globular proteins** that are soluble in aqueous saline solution have their **nonpolar,** hydrophobic R-groups folded to the **inside.** In contrast, their **polar,** hydrophilic R-groups tend to be exposed on the **surface.**

B. **Membrane proteins,** which are in a **nonpolar** environment, have their **hydrophobic** R-groups on the **surface.**

V. PROTEIN DENATURATION. Denaturation of proteins (unfolding into random coils) may result from exposure to a variety of agents.

A. Extremes of pH (e.g., strong acid or alkali)

B. Ionic detergents [e.g., sodium dodecylsulfate (SDS)]

? **C.** Chaotropic agents (e.g., urea, guanidine)

D. Heavy metal ions (e.g., Hg^{++})

E. Organic solvents (e.g., alcohol or acetone)

F. High temperature

G. Surface films (e.g., as when egg whites are beaten)

VI. CLINICAL RELEVANCE

A. Sickle cell anemia. In the **mutant sickle cell hemoglobin (Hgb S),** the hydrophobic **valine** replaces the hydrophilic **glutamate** at position 6 of the β-chain of **normal hemoglobin A (Hgb A).**

1. **Sickle cell disease. Individuals with the homozygous genotype (SS)** have only Hgb S in their red blood cells (RBCs).

    a. **Deoxygenated Hgb S produces fibrous precipitates,** leading to the formation of misshapen RBCs known as **sickle cells.**

    b. **The fragile sickle cells have a shorter life span** than normal RBCs, causing severe anemia.

    c. These dense, inflexible cells may have difficulty passing through the tissue capillaries, resulting in **vaso-occlusion.**

    d. Thus, in addition to anemia, affected patients may have acute episodes of vaso-occlusion **(sickle cell crisis),** with disabling pain that requires hospitalization.

2. **Sickle cell trait. Individuals with the heterozygous genotype (AS)** have both Hgb A and Hgb S in their RBCs.

    a. Patients are usually asymptomatic, with no anemia.

    b. They may have episodes of **hematuria** due to sickling in the renal medulla that is mild and self-limiting.

B. Scurvy. This condition is caused by **defective collagen synthesis** resulting from a **vitamin C (ascorbic acid) deficiency.**

    1. Selected consequences of abnormal collagen in scurvy

        a. Defective would healing

        b. Defective tooth formation

        c. Loosening of teeth

        d. Bleeding gums

        e. Rupture of capillaries

    2. Ascorbic acid is required for the **hydroxylation of proline and lysine** during post-translational processing of collagen.

        a. After the polypeptide chain has been synthesized on the rough endoplasmic reticulum, some of the proline and lysine residues are converted to **hydroxyproline** and **hydroxylysine.**

        b. The hydroxylating reaction requires an enzyme (hydroxylase), O_2, and Fe^{2+}.

        c. Ascorbate is required to maintain the iron in its active oxidation state (Fe^{2+}).

    3. Hydroxyproline forms interchain hydrogen bonds that stabilize the collagen triple helix. The symptoms of scurvy are the result of weakened collagen when these hydrogen bonds are missing.

3

Enzymes

I. ENERGY RELATIONSHIPS

A. Cells need energy to do work, which may involve:

   1. Synthesis

   2. Movement

   3. Transport across membranes

   4. Heat generation

B. Cells obtain energy from chemical reactions. The **free-energy change (ΔG)** is the quantity of energy from these reactions that is available to do work.

II. FREE-ENERGY CHANGE

A. Free-energy change and the equilibrium constant

   1. The ΔG of a reaction $A + B \rightleftharpoons C + D$ is:

$$\Delta G = \Delta G^{0\prime} + RT\ln \frac{[C]\,[D]}{[A]\,[B]}$$

   where $\Delta G^{0\prime}$ is the **standard free-energy change** (when the concentrations of all the reactants and products are 1M and the pH = 7), R is the gas constant (1.987 cal/mol·K), and T is the absolute temperature.

   2. When the reaction has reached **equilibrium**, $[C]\,[D]/[A]\,[B] = K_{eq}$ and **ΔG = 0,** so $\Delta G^{0\prime}$ is related to K_{eq} as follows:

$$\Delta G^{0\prime} = -RT\ln K_{eq}$$

   3. Table 3-1 shows numerical relationships between $\Delta G^{0\prime}$ and K_{eq} at 37°C (310° absolute).

B. Thermodynamic spontaneity

   1. Exergonic reactions, in which K_{eq} **is greater than 1 and $\Delta G^{0\prime}$ is negative,** are **spontaneous** (the reaction goes to the **right** so that the final concentration of the products, C and D, is greater than that of the reactants, A and B).

Table 3–1.
Numerical Relationships
Between $\Delta G^{0'}$ and K_{eq} at 37°C

$\Delta G^{0'}$	K_{eq}
+4255	0.001
+2837	0.01
+1418	0.1
0	1.0
−1418	10.0
−2837	100.0
−4255	1000.0
−7092	100,000.0

2. **Endergonic reactions,** in which K_{eq} **is less than 1 and** $\Delta G^{0'}$ **is positive,** are **nonspontaneous** (the reaction goes to the **left** so that the final concentration of the reactants, A and B, is greater than that of the products, C and D).

3. $\Delta G^{0'}$ **cannot predict spontaneity under intracellular conditions.** Intracellular spontaneity is a function of actual concentrations as well as K_{eq}. ΔG, **not** $\Delta G^{0'}$, is a reflection of intracellular spontaneity.

    a. For example, aldolase, one of the reactants in the breakdown of glucose (glycolysis), has a $\Delta G^{0'}$ of about 5500 cal/mol. The K_{eq} is 0.001. The reaction is nonspontaneous and goes to the left.

    b. If the concentrations of the reactants and products in the aldolase reaction are 0.0001 M (a reasonable intracellular value), the ΔG is −173 cal/mol. The reaction is spontaneous and goes to the right.

C. Enthalpy, entropy, and free-energy change

    1. **Enthalpy.** The enthalpy change (ΔH) is the **amount of heat generated or absorbed** in a reaction.

    2. **Entropy.** The entropy change (ΔS) is a measure of the change in the **randomness** or **disorder** of the system.

        a. An **entropy increase** occurs when a salt crystal dissolves, when a solute diffuses from a more concentrated to a less concentrated solution, and when a protein is denatured.

        b. An **entropy decrease** occurs when a complex molecule is synthesized from smaller substrates.

    3. **Free-energy change** is related to enthalpy and entropy as follows:

$$\Delta G = \Delta H - T\Delta S$$

where T is the absolute temperature (°K).

III. ENZYMES AS BIOLOGICAL CATALYSTS. These molecules control the **rate** of biological reactions.

    A. For a reaction where reactant A is converted to product B (A→B), ΔG of the reactant and product can be plotted against a "reaction coordinate," which represents the course of the reaction under standard conditions (Figure 3-1).

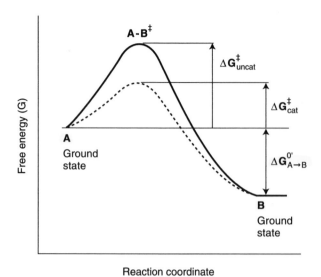

Figure 3-1. The effect of a catalyst on the activation energy of the chemical reaction A → B. The *solid line* represents the reaction in the absence of a catalyst, and the *dotted line*, the reaction in the presence of a catalyst.

B. Direction of reaction

    1. Because catalysts do not change the $\Delta G^{0'}$ they **do not alter the extent or the direction of the reaction.**

    2. If the free energy of the ground state of B is lower than that of A, the **ΔG is negative,** and the reaction proceeds to the right (i.e., toward B).

    3. If, on the other hand, the free energy of the ground state of A is lower than that of B, the **ΔG is positive,** and the reaction proceeds to the left (i.e., toward A).

C. Rate of reaction

    1. The $\Delta G^{0'}$ provides no information concerning the **rate** of conversion from A to B.

    2. When A is converted to B, it first goes through an energy barrier called the **transition state, A-B‡.**

        a. The **activation energy ($\Delta G^{\ddagger}$)** is the energy required to scale the energy barrier and form the transition state.

        b. The **greater the $\Delta G^{\ddagger}$,** the **lower the rate of the reaction** converting A to B.

D. Like other catalysts, enzymes introduce a **new reaction pathway**.

    1. The $\Delta G^{\ddagger}$ is lower.

    2. The **reaction rate** is faster.

IV. MICHAELIS-MENTEN EQUATION. This expression describes the **kinetics** of enzyme reactions.

    A. In enzyme-catalyzed reactions, substrates bind to enzymes at their **active sites,** where conversion to products occurs, followed by the release of **unchanged** enzymes.

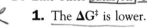

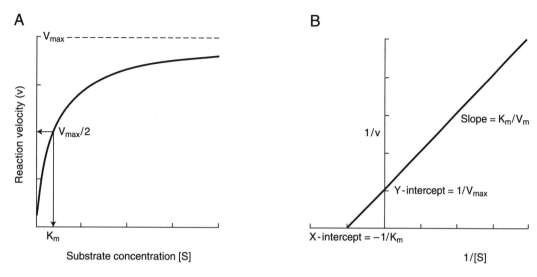

Figure 3-2. The velocity of an enzyme-catalyzed system. (A) Reaction velocity (v) versus substrate concentration ([S]). (B) Lineweaver-Burk (double-reciprocal) plot.

$$E + S \underset{k_2}{\overset{k_1}{\rightleftharpoons}} ES \overset{k_3}{\longrightarrow} E + P$$

where E is the enzyme; S the substrate; ES the enzyme–substrate complex; P the product; and k_1, k_2, and k_3 are rate constants.

B. The ES complex is a transition state with a lower $\Delta G^{\ddagger}$ than the uncatalyzed reaction.

C. The **velocity (v)** of product formation **is related to the concentration of the enzyme–substrate complex:**

$$v = k_3[ES]$$

where k_3 (a rate constant) is also called k_{cat} (particularly in more recent textbooks).

D. The **Michaelis-Menten equation** predicts how velocity is related to substrate concentration if enzyme concentration is held constant:

$$v = \frac{V_m[S]}{K_m + [S]}$$

where V_m is the maximum velocity and K_m, **which equals $(k_2 + k_3)/k_1$,** is the **Michaelis constant.**

E. K_m is the substrate concentration at which $v = 1/2V_m$ ([S] = K_m).

F. A **plot of velocity versus [S]** is a rectangular **hyperbola** (Figure 3-2A).

V. LINEWEAVER-BURK EQUATION. This form of the Michaelis-Menten equation, which is sometimes known as the **double-reciprocal** equation, plots 1/v against 1/[S] to yield a straight line (see Figure 3-2B).

used to find K_m and V_{max}!

$$\frac{1}{v} = \frac{K_m + [S]}{V_m[S]} = \frac{K_m}{V_m} \times \frac{1}{[S]} + \frac{1}{V_m}$$

A. The **slope** is K_m/V_m.

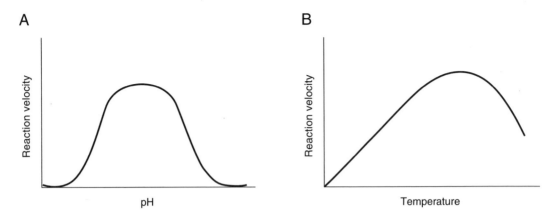

Figure 3-3. Graphic depiction of the effect of pH and temperature on an enzyme-catalyzed reaction. (A) Reaction velocity (v) versus pH. (B) Reaction velocity (v) versus temperature.

   B. The **Y-intercept** is $1/V_m$.

   C. The **X-intercept** is $-1/K_m$.

VI. ENZYME REGULATION

   A. Both pH and temperature affect enzyme activity.

   1. The arms of the v versus pH curve often have the shape of titration curves; this indicates the approximate pKs of groups in the active site (Figure 3-3A).

   2. The v versus temperature curve rises to a maximum and then falls, because denaturation destroys enzymatic activity (see Figure 3-3B).

   B. Inhibitors reduce the activity of enzymes.

   1. Competitive inhibitors are **substrate analogs** that compete with the substrate for the active site of the enzyme.

   a. The apparent K_m **is higher,** but the V_m remains the same.

   b. On a Lineweaver-Burk plot, the **slope is increased, the X-intercept has a smaller absolute value,** and the **Y-intercept is unchanged** (Figure 3-4A).

   2. Noncompetitive inhibitors bind at a site different from the active site.

   a. The V_m is lower, but the K_m is unchanged.

   b. On a Lineweaver-Burk plot, the **slope is increased, the X-intercept is unchanged,** and the **Y-intercept is larger** (see Figure 3-4B).

   3. Uncompetitive inhibitors bind only to the **ES complex.**

   a. Both K_m and V_m are different.

   b. On a Lineweaver-Burk plot, the lines are **parallel.**

   C. Allosteric regulation. A low–molecular-weight **effector** binds to the enzyme at a specific site other than the active site (the **allosteric site**) and alters its activity.

   1. Allosteric enzymes usually have more than one subunit and more than one active site.

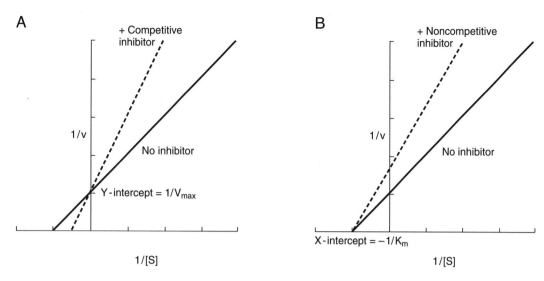

Figure 3-4. Effect of inhibitors on Lineweaver-Burk plots. (A) Effect of a competitive inhibitor. (B) Effect of a noncompetitive inhibitor.

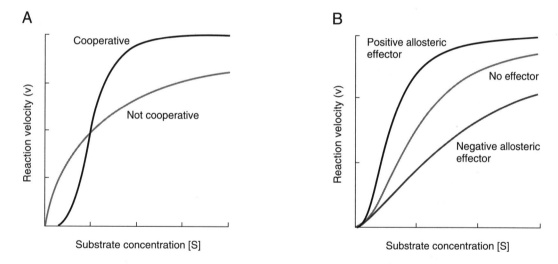

Figure 3-5. Influence of allosteric effectors on allosteric enzymes. (A) Reaction velocity (v) versus substrate concentration ([S]) for enzymes showing cooperative and noncooperative reaction kinetics. (B) Reaction velocity (v) versus substrate concentration ([S]) for an allosteric enzyme showing the effects of negative and positive effectors.

    a. In enzymes with multiple active sites that interact cooperatively, velocity versus [S] curves are **sigmoid** (Figure 3-5A).

    b. The binding of one substrate molecule facilitates the binding of the substrate at other sites

  2. Effectors may have a positive or a negative effect on activity (see Figure 3-5B).

    a. Positive effectors decrease the apparent K_m. [less s required]

      b. **Negative effectors increase** the apparent K_m. *More [S] required*

   3. **Example: muscle hexokinase**

      a. Hexokinase catalyzes the first reaction in the use of glucose by muscle cells:

$$\text{Glucose} + \text{ATP} \rightarrow \text{glucose-6-phosphate} + \text{ADP}$$

      b. Hexokinase has a low K_m (0.1 M) compared to blood glucose concentrations (4–5 mM), so it is **saturated** and **operates at its V_m.**

      c. When glycolysis slows down, glucose-6-phosphate accumulates.

      d. **Glucose-6-phosphate allosterically inhibits hexokinase.**

      e. This keeps the supply of glucose-6-phosphate in balance.

D. **Other mechanisms of enzyme regulation**

   1. Induction or repression of enzyme synthesis by altering gene expression

      a. **Cytochrome P450** enzymes in the liver degrade and **detoxify drugs** (e.g., phenobarbital).

      b. These enzymes are induced by the drugs themselves.

   2. Covalent modification

      a. **Phosphorylase,** the enzyme that breaks down glycogen, is activated by **phosphorylating** a specific hydroxyl group.

      b. This phosphorylation is stimulated by hormones that elevate blood glucose, such as **glucagon** and **epinephrine.**

   3. Protein–protein interaction between an enzyme and a regulatory protein

      a. **Pancreatic lipase,** the enzyme that digests dietary fat, is assisted by **colipase.**

      b. Colipase anchors the lipase to the surface of fat droplets.

VII. CLINICAL RELEVANCE: METHANOL AND ETHYLENE GLYCOL POISONING

A. **Mechanism of poisoning.** Methanol and ethylene glycol toxicity is caused by the action of their metabolites. In both cases, the first oxidation is carried out by alcohol dehydrogenase.

   1. **Methanol** is oxidized to **formaldehyde** and **formic acid.** The eyes are particularly sensitive to **formaldehyde,** so methanol poisoning can quickly lead to blindness.

   2. **Ethylene glycol** is oxidized to **glycoaldehyde, oxalate,** and **lactate. Kidney failure** due to deposition of **oxalate crystals** is a frequent consequence of ethylene glycol poisoning.

B. **Treatment** involves an initial infusion of ethanol.

   1. Ethanol is a **competitive substrate**, and displaces methanol or ethylene glycol from the active site of alcohol dehydrogenase.

   2. This measure prevents continued production of toxic metabolites.

4

Citric Acid Cycle and Oxidative Phosphorylation

I. CELLULAR ENERGY AND ADENOSINE TRIPHOSPHATE

A. Coupled chemical reactions provide the energy needed for cellular work. Energy-rich reduced fuel molecules (from food) are oxidized in **catabolic reaction sequences** called **pathways**. Catabolic reactions are coupled to reactions that result in the combination of **adenosine diphosphate (ADP)** with **inorganic phosphate (P$_i$)** to form **adenosine triphosphate (ATP)** [Figure 4-1].

B. High-energy phosphate compounds are frequently involved in driving otherwise unfavorable (endergonic) reactions.

   1. High-energy bonds, which are very **reactive,** have a **high free energy ($\Delta G^{0\prime}$) of hydrolysis** that ranges between −7 and −15 kcal/mol.

   2. Energy level of phosphate bonds in ATP (see Figure 4-1)

   a. Phosphoanhydride bonds are high-energy bonds.

   b. Phosphate ester bonds are low-energy bonds.

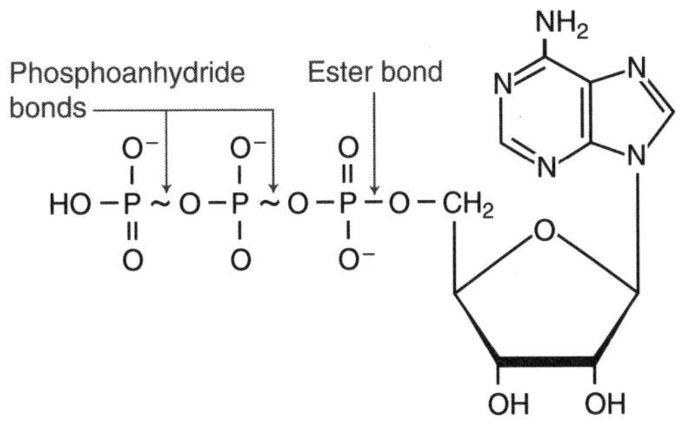

Figure 4-1. Structure of adenosine triphosphate (ATP), showing the location of high-energy phosphoanhydride bonds and low-energy phosphate ester bonds. ~ = high-energy bonds.

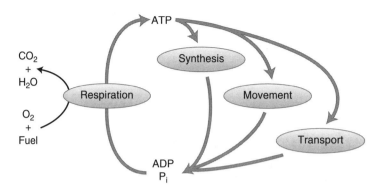

Figure 4-2.   The high-energy phosphate cycle.

C. In a biochemical pathway, a particular exergonic (favorable) reaction can provide the energy for a particular endergonic (unfavorable) reaction if the two reactions have a **common intermediate** [e.g., the formation of oxaloacetate (OAA) is coupled to synthesis of citric acid].

D. Other sources of energy

   1. Change in protein conformation (e.g., ATP synthesis by mitochondria).

   2. Flow of ions across membranes (e.g., rotation of bacterial flagella).

E. Synthetic reactions, Motion and transport across membranes are coupled to reactions that use ATP and release ADP and P_i (Figure 4-2).

II. CITRIC ACID CYCLE (Figure 4-3)

A. The cycle is located in the mitochondria. These organelles occur in all body cells, except for red blood cells.

B. The citric acid cycle (tricarboxylic acid cycle, Krebs cycle) is the **final common pathway** of oxidative metabolism.

C. Acetyl coenzyme A (acetyl CoA) condenses with OAA to begin the cycle. Catabolism of carbohydrates, fats, and proteins provides the acetyl CoA.

   1. Glucose catabolism eventually produces pyruvate, which yields acetyl CoA via pyruvate dehydrogenase.

   2. Fatty acids generate acetyl CoA via β-oxidation.

   3. Some amino acids are degraded to acetyl CoA.

III. PRODUCTS OF THE CITRIC ACID CYCLE (ONE REVOLUTION)

A. Release of **two moles of CO_2** and regeneration of **one mole of OAA** for oxidation of one acetyl CoA; most of the CO_2 from body metabolism is produced this way.

B. Generation of 11 molecules of ATP via oxidative phosphorylation.

C. Production of one equivalent of high-energy phosphate as guanosine triphosphate (GTP) [or ATP].

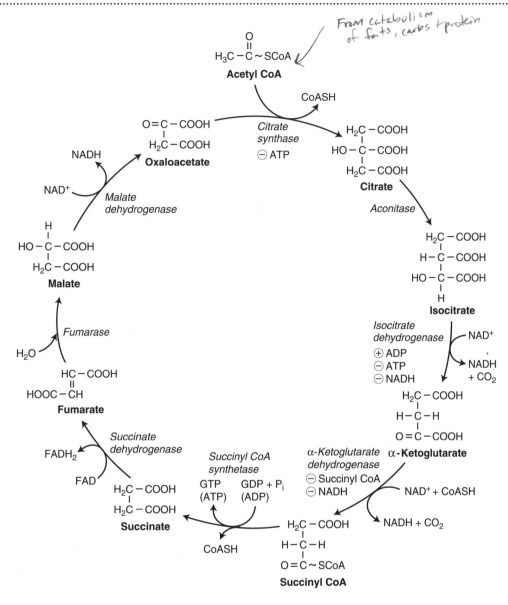

Figure 4-3. The citric acid cycle (tricarboxylic acid cycle, Krebs cycle). + = activator; − = inhibitor; *italicized terms* = enzyme names; ~ = high-energy compounds.

IV. SYNTHETIC FUNCTION OF THE CITRIC ACID CYCLE

A. **Intermediates** also serve as substrates for biosynthetic pathways and thus need to be replenished.

B. **Anaplerotic reactions** provide OAA or other cycle intermediates.

1. **Pyruvate carboxylase** in the liver and kidney.

$$\text{Pyruvate} + \text{ATP} + \text{HCO}_3^- \rightleftharpoons \text{OAA} + \text{ADP} + \text{P}_i$$

2. **Phosphoenolpyruvate (PEP) carboxykinase** in the heart and skeletal muscle.

$$\text{Phosphoenolpyruvate} + CO_2 + GDP \rightleftharpoons OAA + GTP$$

3. **Malic enzyme** in many tissues.

$$\text{Pyruvate} + HCO_3^- + NAD(P) \rightleftharpoons \text{Malate} + NAD(P)^+$$

4. **Glutamate dehydrogenase** in the liver.

$$\text{Glutamate} + NAD(P)^+ + H_2O \rightleftharpoons \alpha\text{-ketoglutarate} + NAD(P)H + NH_4^+$$

V. REGULATION OF THE CITRIC ACID CYCLE. Control of the flow of substrates through the cycle occurs at three highly exergonic steps (see Figure 4-3):

A. **Acetyl CoA** condenses with **OAA** to form **citrate.**

   1. Enzyme: **citrate synthase.**

   2. **ATP,** an **allosteric inhibitor,** increases the K_m for acetyl CoA, one of the substrates.

B. **Isocitrate** is oxidized to α-**ketoglutarate.**

   1. Enzyme: **isocitrate dehydrogenase.**

   2. **ADP** is an allosteric **activator,** and **ATP** and **NADH** are **inhibitors.**

C. α-**Ketoglutarate** is converted to **succinyl CoA.**

   1. Enzyme: α-**ketoglutarate dehydrogenase.**

   2. **Succinyl CoA** and **NADH** are **inhibitors.**

VI. ELECTRON TRANSPORT AND OXIDATIVE PHOSPHORYLATION

A. Each turn of the citric acid cycle generates three NADH and one $FADH_2$.

B. In the mitochondrial **electron transport system (ETS)** electrons pass from NADH or $FADH_2$ to ultimately reduce O_2 and produce H_2O (Figure 4-4).

C. Most of the ATP in aerobic cells is generated by mitochondrial **oxidative phosphorylation,** which uses the energy derived from the flow of electrons through the ETS to drive the **synthesis of ATP** from ADP and P_i.

D. The **oxidation–reduction potential** of electrons depends on the compound to which they once belonged.

   1. The oxidation–reduction potential is directly related to the ΔG:

$$E = E_o' + \frac{2.3RT}{nF} \log\left(\frac{[\text{Oxidant}]}{[\text{Reductant}]}\right)$$

$$\Delta G^{o\prime} = -nF\Delta E_o'$$

where E = potential in volts, F = Faraday's constant, and n = the number of electrons.

   2. The difference in oxidation–reduction potential between NADH and O_2 (−52.6 kcal/mol) or $FADH_2$ and O_2 (−48 kcal/mol) is sufficient to drive the synthesis of ATP from ADP and P_i (+7.3 kcal/mol) several times.

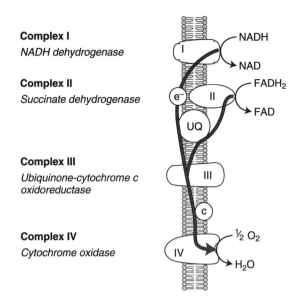

Complex I
NADH dehydrogenase

Complex II
Succinate dehydrogenase

Complex III
Ubiquinone-cytochrome c oxidoreductase

Complex IV
Cytochrome oxidase

Figure 4-4. Diagrammatic representation of the mitochondrial electron transport system (ETS). The path of the electrons is indicated by the broad, shaded arrows. UQ = ubiquinone; C = cytochrome c.

VII. CHEMIOSMOTIC HYPOTHESIS. This hypothesis describes the coupling of electron flow through the ETS to ATP synthesis (Figure 4-5).

  A. Respiratory complexes as proton pumps. As electrons (e^-) pass through complexes I, III, and IV, hydrogen ions (H^+) are "pumped" across the inner mitochondrial membrane into the intermembrane space.

    1. The H^+ concentration in the intermembrane space increases relative to the mitochondrial matrix.

    2. This generates a **proton-motive force** as a result of two factors:

      a. Difference in pH (ΔpH).

      b. **Difference in electrical potential** ($\Delta\psi$) between the intermembrane space and the mitochondrial matrix.

  B. ATP synthase complex (complex V). Hydrogen ions pass back into the matrix through complex V, and in doing so, drive the **synthesis of ATP.**

    1. Passage of a pair of electrons from **NADH** through the ETS to O_2 generates **three ATP.**

    2. The passage of a pair of electrons from **FADH$_2$** to O_2, which bypasses complex I, generates **two ATP.**

VIII. CLINICAL RELEVANCE

  A. Uncoupling agents

    1. These substances carry H^+ across the inner mitochondrial membrane without going through complex V. This short-circuits the proton gradient and **uncouples electron flow from ATP synthesis.** The **energy** that would have been used to synthesize ATP is **dissipated as heat.**

    2. Dinitrophenol and some other hydrophobic organic acids that can carry protons across the inner mitochondrial membrane are **chemical uncoupling agents.** Dinitrophenol was formerly used as a medication for weight reduction. It caused blindness in some patients (the retina has a very high rate of oxidative metabolism).

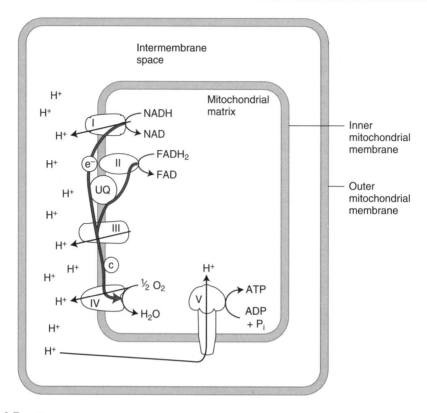

Figure 4-5.   Diagrammatic representation of the Mitchell chemiosmotic hypothesis of oxidative phosphorylation.

Table 4-1.
Inhibitors of Electron Transport

Site	Inhibitors
Complex I	Amobarbital (barbiturate)
	Rotenone (insecticide)
	Piericidin A (antibiotic)
Complex II	Antimycin A (antibiotic)
Complex IV	Cyanide
	Hydrogen sulfide
	Carbon monoxide

   3. The mitochondria of **brown fat** in newborn mammals, including humans, contain **thermogenin (uncoupling protein, UCP),** which allows protons to pass through the inner membrane without synthesizing ATP. The **energy** is **dissipated as heat** to maintain normal body temperature.

B. Inhibitors block electron flow through one of the complexes (Table 4-1). That is why H_2S and HCN are such lethal poisons.

5

Carbohydrate Metabolism

I. CARBOHYDRATE DIGESTION AND ABSORPTION

A. Dietary carbohydrate is **digested in the mouth and intestine** and **absorbed from the small intestine.**

B. Disaccharides (e.g., sucrose, lactose), **oligosaccharides** (e.g., dextrins), and **polysaccharides** (e.g., starch) are cleaved into monosaccharides (e.g., glucose, fructose).

    1. **Starch,** the storage form of carbohydrate in plants, is hydrolyzed to maltose, maltotriose, and α-limit dextrins by **the enzyme α-amylase in saliva and pancreatic juice.**

    2. **Disaccharides and oligosaccharides** are hydrolyzed to monosaccharides by **enzymes on the surface of epithelial cells in the small intestine.**

C. **Monosaccharides** are absorbed directly by carrier-mediated transport. These sugars (primarily glucose) travel **via the portal vein** to the **liver** for:

    1. Oxidation to CO_2 and H_2O for energy

    2. Storage as glycogen

    3. Conversion to triglyceride (fat)

    4. Release into the general circulation (as glucose)

II. GLYCOGEN METABOLISM. Glycogen, the storage form of carbohydrate in the human body, is found chiefly in the liver and muscle (Figure 5-1).

A. Glycogenesis (glycogen synthesis)

    1. **Uridine diphosphate-glucose** is the activated substrate.

    2. The enzyme **glycogen synthase** adds glucosyl units to the nonreducing ends of existing chains in **α-1,4 linkages.**

    3. The branching enzyme **amylo (1→4) to (1→6) transglycosylase** moves pieces that contain about seven glucose residues from the nonreducing ends of the chains to the interior and creates branches with **α-1,6 linkages.**

B. Glycogenolysis (glycogen breakdown)

    1. The enzyme **phosphorylase** releases units of **glucose 1-phosphate** from the nonreducing ends one at a time.

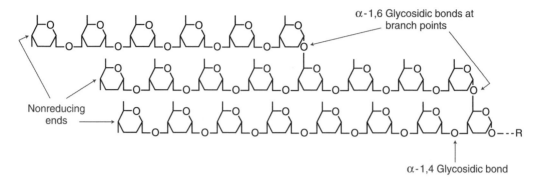

Figure 5-1. Glycogen, a polymer of glucose units linked by α-1,4-glycosidic bonds with α-1,6-glycosidic bonds at the branch points.

    2. The enzyme **phosphoglucomutase** converts the glucose 1-phosphate to glucose 6-phosphate.

    3. A **debranching** system releases **glucose residues** from the α-1,6 bonds at the branch points.

  C. **Regulation of glycogenesis and glycogenolysis.**

    1. **Glucagon** (acting on liver) and **epinephrine** (acting on muscle and liver) **stimulate glycogenolysis** and **inhibit glycogenesis** via the cyclic adenosine monophosphate (cAMP) protein kinase A phosphorylation cascade.

    2. **Insulin stimulates glycogenesis** and **inhibits glycogenolysis** via dephosphorylation in muscle, liver, and adipose tissue.

III. **GLYCOLYSIS.** The biochemical pathway known as glycolysis involves the oxidation of glucose (Figure 5-2).

  A. Glycolysis occurs in the **cytosol** in most tissues of the body.

  B. Under **anaerobic** conditions (without oxygen), glycolysis involves the conversion of glucose to lactate.

$$\text{Glucose} \rightarrow 2\text{Lactate} + 2\text{ATP}$$

    1. Anaerobic glycolysis is characteristic of skeletal muscle after prolonged exercise.

    2. The enzyme **lactate dehydrogenase** converts pyruvate to lactate. The NADH that is produced by glycolysis becomes NAD^+. No additional ATP is generated (Figure 5-3).

  C. Under **aerobic** conditions (with oxygen), glycolysis converts glucose to CO_2 and H_2O.

$$\text{Glucose} + 6O_2 \rightarrow 6CO_2 + 6H_2O + 36\text{–}38 \text{ ATP}$$

    1. Aerobic glycolysis is characteristic of the brain.

    2. The NADH is oxidized by the **mitochondrial electron transport system.** ATP is generated by **oxidative phosphorylation.**

  D. **Phosphorylation,** the first step in glycolysis, involves the reaction of glucose in the presence of **hexokinase** or **glucokinase** to form **glucose 6-phosphate** (see Figure 5-2).

    1. **Hexokinase** is found in the cytosol of most tissues. It has several properties:

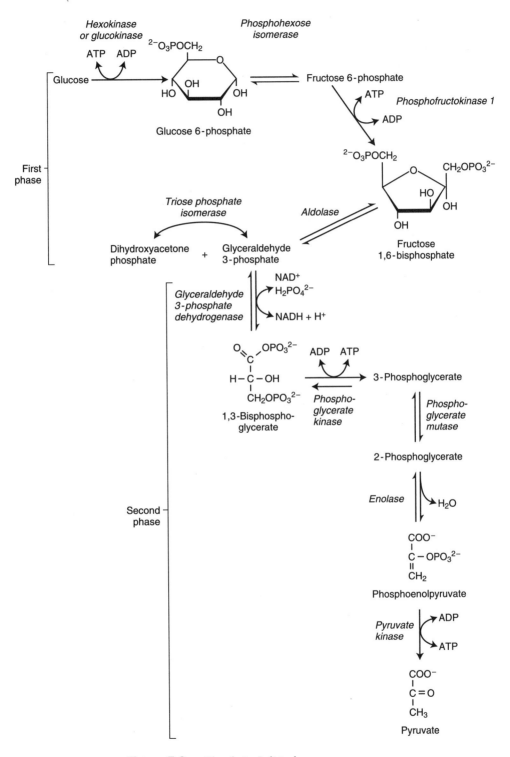

Figure 5-2. Glycolysis. *Italicized terms* = enzyme names.

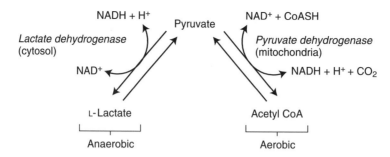

Figure 5-3. Fates of pyruvate in anaerobic versus aerobic conditions. *Italicized terms* = enzyme names.

    a. Low specificity, which means that it can catalyze the phosphorylation of a wide variety of hexoses.

    b. Low K_m, which means that it is saturated at normal blood glucose concentrations.

    c. Inhibition by **glucose 6-phosphate,** which prevents cells from accumulating too much glucose (phosphorylation traps glucose inside cells).

  2. Glucokinase (hexokinase D) is present in the **liver** and **pancreas** (β-cells). It has several properties:

    a. High specificity for glucose.

    b. High K_m (above the normal blood glucose concentration).

    c. Inhibition by **fructose 6-phosphate,** which ensures that glucose will be phosphorylated only as fast as it is metabolized.

 C. In the first phase of glycolysis (five reactions), one mole of glucose is converted to two moles of glyceraldehyde 3-phosphate. **Two moles of ATP** are **consumed** for each mole of glucose.

 D. In the second phase (five reactions), the two moles of glyceraldehyde 3-phosphate are oxidized to two moles of pyruvate. **Four moles of ATP** and **two moles of NADH** are generated for each mole of glucose.

 E. NADH itself does not pass through the mitochondrial inner membrane.

  1. The **glycerol phosphate shuttle** (most tissues) transfers electrons from cytosolic NADH to mitochondrial $FADH_2$. It generates **two moles of ATP per mole of cytosolic NADH** or **36 moles of ATP per mole of glucose** oxidized.

  2. The **malate–aspartate shuttle** (heart, muscle, and liver) transfers electrons to mitochondrial NADH. It generates **three moles of ATP per mole of cytosolic NADH** or **38 moles of ATP per mole of glucose.**

IV. GLUCONEOGENESIS. This process, which occurs primarily in the liver and kidney, is the synthesis of glucose from small noncarbohydrate precursors such as lactate and alanine.

 A. Gluconeogenesis involves the reversible reactions of glycolysis. To bypass the nonreversible steps of glycolysis, separate reactions occur.

  1. Conversion of **pyruvate** to **phosphoenolpyruvate (PEP)** [Figure 5-4] bypasses pyruvate kinase.

  2. Conversion of **fructose 1,6-biphosphate** to **fructose 6-phosphate** by **fructose 1,6-bisphosphatase** bypasses phosphofructokinase.

  3. Conversion of **glucose 6-phosphate** to **glucose** by **glucose 6-phosphatase** bypasses hexokinase.

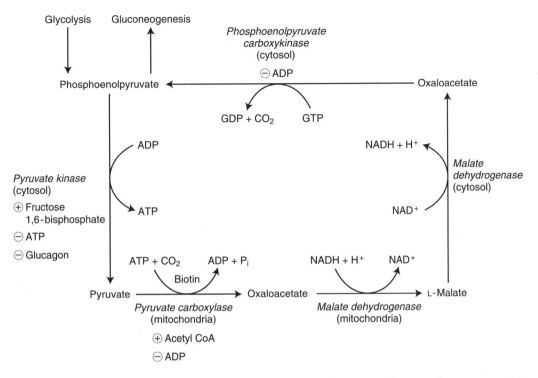

Glycolysis   Gluconeogenesis

Phosphoenolpyruvate carboxykinase (cytosol)

$\ominus$ ADP

Phosphoenolpyruvate ← ← Oxaloacetate

GDP + CO_2 GTP

ADP

NADH + H^+

Malate dehydrogenase (cytosol)

Pyruvate kinase (cytosol)

$\oplus$ Fructose 1,6-bisphosphate

$\ominus$ ATP

$\ominus$ Glucagon

ATP

NAD^+

ATP + CO_2 ADP + P_i NADH + H^+ NAD^+

Biotin

Pyruvate → Oxaloacetate → L-Malate

Pyruvate carboxylase (mitochondria)

$\oplus$ Acetyl CoA

$\ominus$ ADP

Malate dehydrogenase (mitochondria)

Figure 5-4. The pathways between pyruvate and phosphoenolpyruvate. Glucagon elevates intracellular cAMP, which stimulates protein kinase A phosphorylation and inactivation of pyruvate kinase. *Italicized terms* = enzyme names; $\oplus$ = stimulation; $\ominus$ = inhibition.

B. Glucose from gluconeogenesis is released into the bloodstream for transport to tissues such as the brain and exercising muscle.

C. Gluconeogenic substrates

   1. Lactate

   2. Pyruvate

   3. Glycerol

   4. Substances that can be converted to oxalacetate via the citric acid cycle (such as amino acid carbon skeletons)

D. The **Cori cycle** describes the shuttling of gluconeogenic substrates between muscle and the liver.

   1. Lactate from exercising or ischemic muscle is carried by the circulation to the liver and serves as a substrate for gluconeogenesis.

   2. The liver releases the resynthesized glucose into the circulation for transport back to the muscle.

V. REGULATION OF GLYCOLYSIS AND GLUCONEOGENESIS

   A. This is accomplished by control of the magnitude and direction of flow at two steps:

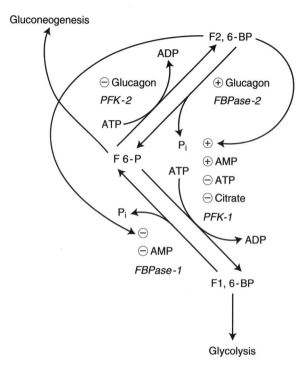

Figure 5-5.   Regulation of phosphofructokinase and fructose 6-phosphatase in liver cells. Glucagon elevates intracellular cAMP, which stimulates protein kinase A phosphorylation of the bifunctional enzyme phosphofructokinase-2/fructose bisphosphatase-2 (PFK-2/FBPase-2). Phosphorylation elevates FBPase-2 activity and depresses PFK-2 activity. In turn, the resulting fall in fructose 2,6-bisphosphate (F2, 6-BP) inactivates PFK-1 and activates FBPase-1, leading to increased flow of F2, 6-BP to fructose 6-phosphate, thus favoring gluconeogenesis over glycolysis. *Italicized terms* = enzyme names; ⊕ = stimulation; ⊖ = inhibition.

>   **1.** Between **fructose 6-phosphate** and **fructose 1,6-bisphosphate**. The activities of phosphofructokinase-1 and fructose 1,6-bisphosphatase are regulated by the supply of adenine nucleotides, citrate, and fructose 2,6-bisphosphate (Figure 5-5).

>   **2.** Between **PEP** and **pyruvate** (see Figure 5-4).

>   **B.** During starvation, blood glucose is low and glucagon is secreted, favoring gluconeogenesis in the liver.

>   **C.** In the fed (absorptive) state, blood glucose is high and glucagon secretion is suppressed, favoring glycolysis.

VI. PENTOSE PHOSPHATE PATHWAY. This pathway may function as an alternate form of glycolysis or may be the route for the complete oxidation of glucose (it begins with glucose 6-phosphate) [Figure 5-6].

>   **A.** The **irreversible oxidative portion generates NADPH,** which is needed for biosynthetic pathways such as fatty acid and cholesterol synthesis.

>   **B.** The **reversible nonoxidative portion** rearranges the sugars so they can reenter the glycolytic pathway.

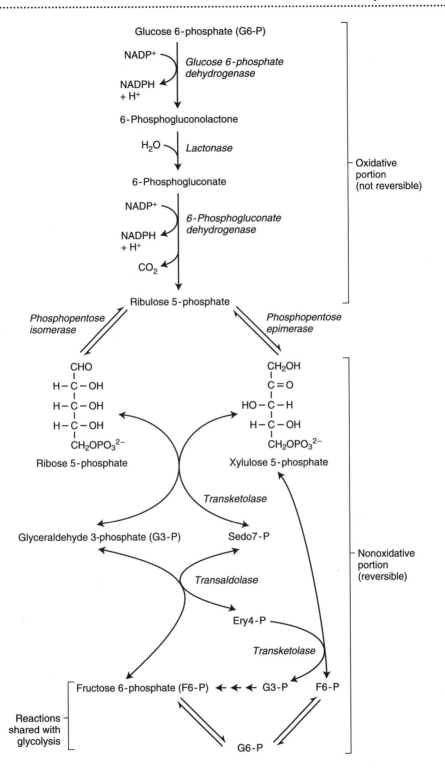

Figure 5-6. The pentose phosphate pathway. Sedo7-P = sedoheptulose 7-phosphate; ery4-P = erythrose 4-phosphate; *italicized terms* = enzyme names.

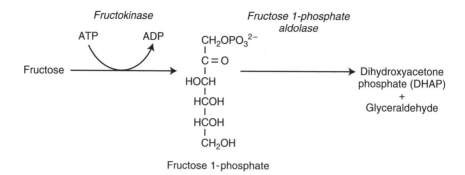

Figure 5-7.   The liver pathway for fructose entry into glycolysis. *Italicized terms* = enzyme names.

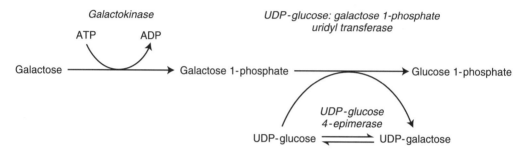

Figure 5-8.   The pathway for converting galactose to glucose 1-phosphate. *Italicized terms* = enzyme names.

   C. Ribose 5-phosphate, which is needed for nucleotide synthesis, can be formed from glucose 6-phosphate by either arm.

VII. SUCROSE AND LACTOSE METABOLISM

   A. Sucrose (cane sugar) and lactose (milk sugar), the common dietary disaccharides, are digested in the small intestine and appear in the circulation as monosaccharides. Some monosaccharides have specialized metabolic pathways.

   B. The enzyme **sucrase** converts sucrose to **glucose** and **fructose.**
   1. The enzyme **hexokinase** can convert **fructose to fructose 6-phosphate** via phosphorylation in muscle and kidney.
   2. Fructose enters glycolysis by a different route in the liver (Figure 5-7).
   3. Dihydroxyacetone phosphate (DHAP) enters glycolysis directly.
   4. After glyceraldehyde is reduced to glycerol, it is phosphorylated and then reoxidized to DHAP.

   C. The enzyme **lactase** converts **lactose** to **glucose and galactose.**
   1. The enzyme **galactokinase** catalyzes the reaction of **galactose to galactose 1-phosphate** via phosphorylation.
   2. In a series of reactions, galactose 1-phosphate becomes glucose 1-phosphate (Figure 5-8).

Table 5-1.
Clinical Effects of Glycogen Storage Diseases

Name and Type of Disease	Enzyme Defect	Tissue	Glycogen in Affected Cells	Clinical Manifestation
Von Gierke's (type I)	Glucose 6-phosphatase	Liver and kidney	Increased amount; normal structure	Hepatomegaly, failure to thrive, hypoglycemia, ketosis, hyperuricemia, hyperlipidemia
Pompe's (type II)	α-1,4-glucosidase	Lysosomes, all organs	Increased amount; normal structure	Failure of cardiac and respiratory systems, death before 2 years of age
Cori's (type III)	Debranching enzyme	Muscle and liver	Increased amount; short outer branches	Similar to type I, but milder
Anderson's (type IV)	Branching enzyme	Liver and spleen	Normal amount; very long outer branches	Liver cirrhosis, death before 2 years of age
McArdle's (type V)	Phosphorylase	Muscle	Moderate increase in amount; normal structure	Painful muscle cramps with exercise
Hers' (type VI)	Phosphorylase	Liver	Increased amount	Similar to type I, but milder
Type VII	Phosphofructokinase	Muscle	Increased amount; normal structure	Similar to type V
Type VIII	Phosphorylase kinase	Liver	Increased amount; normal structure	Mild hepatomegaly, mild hypoglycemia

VIII. CLINICAL RELEVANCE

A. **Glycogen storage diseases** are inherited enzyme deficiencies (Table 5-1).

B. **Hereditary enzyme deficiences in sucrose metabolism**

  1. **Fructokinase deficiency** leads to **essential fructosuria**, a benign disorder.

  2. Some individuals have a **fructose 1-phosphate aldolase deficiency,** which leads to **hereditary fructose intolerance,** characterized by severe hypoglycemia after ingesting fructose (or sucrose).

C. Inherited enzyme deficiencies in lactose metabolism

  1. **Lactase deficiency** sometimes develops in adult life and leads to **milk intolerance** with bloating, flatulence, and diarrhea.

  2. **Galactokinase deficiency** causes a mild form of **galactosemia,** with early cataract formation.

  3. **Galactose 1-phosphate uridyl tranferase deficiency** causes a severe form of galactosemia with growth failure, mental retardation, and even early death.

6

Lipid Metabolism

I. LIPID FUNCTION

A. Fat (triacylglycerol, TG)

    1. **Major fuel store** of the body

    2. **Padding** to protect delicate tissues (e.g., eye, kidney) against trauma

    3. **Insulation** against heat loss

B. **Phospholipids.** These substances are **key components of biological membranes** and of the **lipoproteins** that transport lipids in blood.

C. **Sphingolipids** are also components of membranes.

D. Cholesterol

    1. **Key component of membranes**

    2. **Precursor of bile acids, bile salts,** and **several hormones** (e.g., adrenal corticosteroids, sex steroids, calcitriol)

II. LIPID DIGESTION (Figure 6-1)

A. **Digestion.** Because lipids are **water insoluble,** they must be **emulsified** so the enzymes from the aqueous phase can digest them.

    1. In the **mouth,** medium-chain TGs are first hydrolyzed by the enzyme lipase. This process continues in the **stomach,** producing a mixture of diacylglycerols and free fatty acids (FFAs).

    2. In the **duodenum,** dietary lipids are **emulsified** by **bile salts,** which are synthesized from cholesterol in the liver.

    3. In the **small intestine,** the emulsified fats are **hydrolyzed** by **pancreatic lipase,** phospholipids by **phospholipase A,** and cholesterol esters by a **cholesterol esterase.**

    4. Mixed **micelles** form, which contain fatty acids; diacylglycerols; monoacylglycerols; phospholipids; cholesterol; vitamins A, D, E, and K (ADEK); and bile acids.

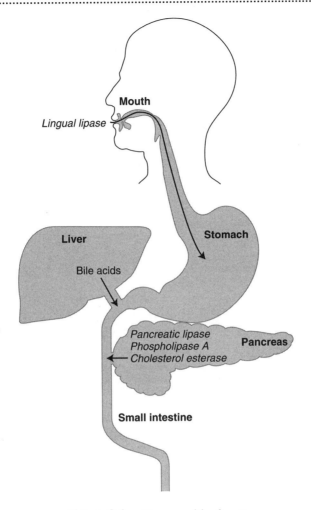

Figure 6-1. Cartoon of fat digestion.

5. The **micelles** are **absorbed** into the cells of the microvilli of the **small intestine,** where they are further metabolized; the products are transported into the circulation.

   a. Medium-chain TGs are hydrolyzed.

   b. Medium-chain fatty acids (MCFAs, 8–10 carbons) pass into the portal vein blood.

   c. Long-chain fatty acids (LCFAs, > 12 carbons) are reincorporated into TG.

   d. The TGs are incorporated into **chylomicrons,** which pass into the **lymphatics** and enter the circulation via the **thoracic duct.**

B. Transport. Lipids are transported to the tissues in the blood plasma primarily as **lipoproteins,** which are spherical particles with a core that contain varying proportions of hydrophobic triacylglycerols and cholesterol esters with an outer layer of cholesterol, phospholipids, and specific **apoproteins.**

III. LIPOPROTEIN ABSORPTION

A. Exogenous lipid (from the intestine), except for MCFAs, is released into the plasma as **chylomicrons.**

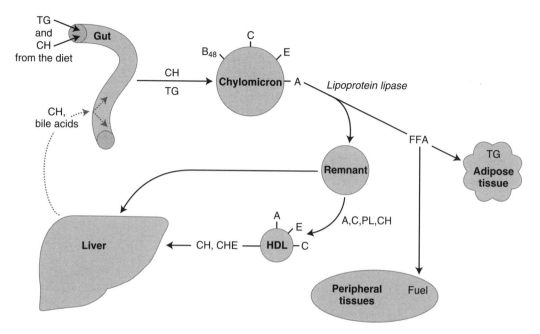

Figure 6-2. Transport of exogenous lipids in the blood. *A, B₄₈, C, E* = apoproteins A, B_{48}, C, E; *CH* = cholesterol; *CHE* = cholesterol esters; *FFA* = free fatty acid; *HDL* = high-density lipoprotein; *PL* = phospholipid; *TG* = triacylglycerol.

   1. Chylomicrons, the largest and least dense of the plasma lipoproteins, contain a high proportion of TGs.
   2. Chylomicron TG is hydrolyzed to **FFAs** and glycerol by **lipoprotein lipase** on the surface of capillary endothelium in **muscle** and **adipose tissue** (Figure 6-2).
   3. The cholesterol-rich **chylomicron remnants** travel to the liver, where they are taken up by **receptor-mediated endocytosis** (RME). They are degraded in the lysosomes.

B. Some **FFAs** are released by adipose tissue into the circulation, and they are absorbed by muscle cells for oxidation. Other FFAs may be stored in adipose tissue.
   1. FFAs may be bound to serum albumin, in which case they are called **nonesterified fatty acids,** and transported to other tissues.
   2. Adipose tissue triacylglycerol is hydrolyzed by hormone-sensitive **lipase** to FFA and glycerol. The lipase is activated by **glucagon** and **epinephrine** via the adenyl cyclase-cAMP-protein kinase A cascade.

C. Endogenous lipid (from the liver) is released into the blood as **very-low-density lipoprotein** (VLDL) [Figure 6-3].
   1. VLDL triglyceride is hydrolyzed by the enzyme **lipoprotein lipase** to FFAs and glycerol, yielding **low-density lipoproteins** (LDLs).
   2. LDLs are removed from the circulation by RME in tissues that contain **LDL receptors,** in part by peripheral tissues that need the cholesterol, but mostly by the **liver.**
      a. LDL cholesterol inhibits **HMG coenzyme A (CoA) reductase,** the rate-limiting step in cellular cholesterol synthesis.

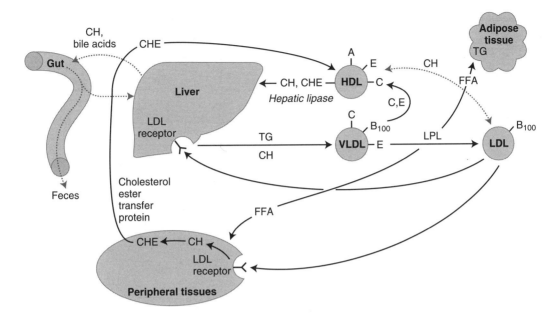

Figure 6-3. Transport of endogenous lipids in the blood. *A*, *B$_{100}$*, *C*, *E* = apoproteins A, B$_{100}$, C, E; *CH* = cholesterol; *CHE* = cholesterol esters; *HDL* = high-density lipoprotein; *LDL* = low-density lipoprotein; *LPL* = lipoprotein lipase; *TG* = triacylglycerol; *VLDL* = very-low-density lipoprotein.

        b. LDL cholesterol **down-regulates** LDL receptor synthesis, in turn causing a decrease in LDL uptake.

        3. High-density lipoproteins (HDLs) are synthesized by the liver. They exchange apoproteins and lipids between plasma lipoprotein particles and also serve in **reverse cholesterol transport.**

IV. OXIDATION OF FATTY ACIDS

    A. Fatty acids are oxidized in the **mitochondrial matrix.** The overall process is:

$$\underset{\text{Fatty acids}}{RCH_2CH_2COOH} \xrightarrow{\beta\text{-oxidation}} \underset{\text{Acetyl CoA}}{CH_3COSCoA} \xrightarrow{\text{Citric acid cycle}} CO_2 + H_2O$$

    B. Fatty acids must first be activated as their acyl CoA thioesters (Figure 6-4).

        1. LCFAs (> 12) are activated in the cytosol.

        2. Long-chain acyl CoAs cannot cross the mitochondrial inner membrane. They are shuttled into the matrix by the **carnitine** transport system (see Figure 6-4).

        3. MCFAs (< 12) pass directly into the mitochondria and are activated in the matrix.

    C. Fatty acyl CoA is oxidized to CO_2 and H_2O by the mitochondrial **β-oxidation system** (Figure 6-5).

        1. β-oxidation proceeds in a repetitive cycle until the fatty acid moiety has been completely converted to acetyl CoA.

        2. Each cycle of β-oxidation generates 5 ATP via the electron transport system and 12 ATP via the combined action of the citric acid cycle and the electron transport system.

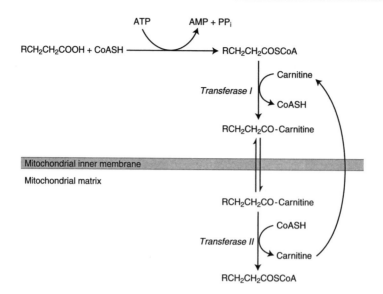

Figure 6-4. Fatty acid activation and the carnitine shuttle for transport of long-chain fatty acids into the mitochondrial matrix. *Italicized terms* = enzyme names.

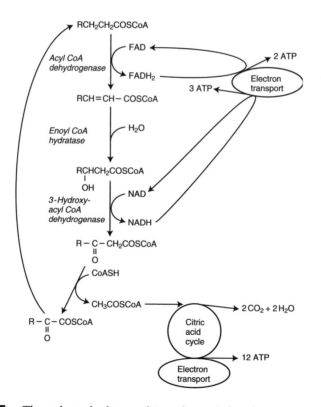

Figure 6-5. The pathway for fatty acid β-oxidation. *Italicized terms* = enzyme names.

3. The terminal three carbons of **odd-numbered fatty acids** yield **propionyl CoA** as the final product of β-oxidation.

  a. Propionyl CoA, which can be carboxylated to succinyl CoA in a three-reaction sequence requiring biotin and vitamin B_{12}, then enters the citric acid cycle.

  b. Propionyl CoA can be used for gluconeogenesis.

D. **Ketogenesis.** Some of the acetyl CoA from β-oxidation is metabolized to **acetoacetate** and **β-hydroxybutyrate** in the **liver.**

  1. Acetyl CoA reacts with acetoacetyl CoA, forming **hydroxymethylglutaryl CoA (HMG CoA).**

  2. HMG CoA then splits to yield **acetoacetate** and acetyl CoA.

  3. Acetoacetate may be reduced by NADH to **β-hydroxybutyrate**, and some of the acetoacetate spontaneously decarboxylates to acetone.

  4. **Extrahepatic tissues,** especially heart muscle, can activate acetoacetate at the expense of succinyl CoA and burn the acetoacetyl CoA for energy.

  5. **Glucose-starved brain** can use acetoacetate for fuel, because this substance is freely soluble in blood and easily crosses the blood–brain barrier.

V. FATTY ACID SYNTHESIS

A. This process is carried out by **fatty acid synthase,** a multienzyme complex in the cytosol. The **primary substrates** are **acetyl CoA** and **malonyl CoA** (Figure 6-6).

B. **Acetyl CoA** is formed in the mitochondria, principally by the enzyme **pyruvate dehydrogenase.**

  1. Acetyl CoA is transported from mitochondria to cytosol by the **citrate-malate-pyruvate shuttle** (Figure 6-7).

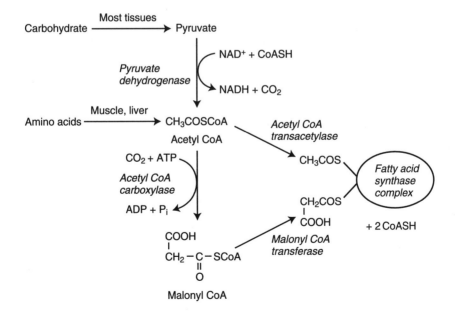

Figure 6-6. Origin of the substrates for fatty acid synthesis. *Italicized terms* = enzyme names.

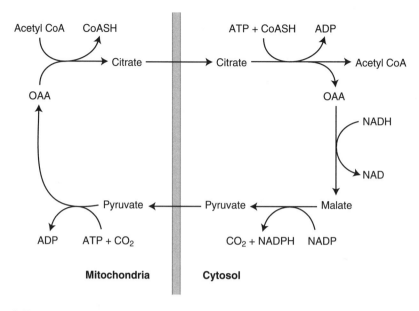

Figure 6-7. The citrate shuttle for transport of acetyl CoA from the mitochondrion to the cytosol.

2. The electrons from one NADH are transferred to NADPH, which is then available for the reductive steps of fatty acid synthesis. NADPH is also supplied by the pentose phosphate cycle.

C. **Malonyl CoA** is formed by the biotin-linked carboxylation of acetyl CoA (see Figure 6-6).

D. The **acetyl and malonyl moieties** are transferred from the sulfur of CoA to active **sulfhydryl groups** in the **fatty acid synthase** (see Figure 6-6), where the synthetic sequence takes place (Figure 6-8).

  1. Enzyme activities in the complex carry out condensation, reduction, dehydration, and reduction.

  2. Seven cycles lead to production of palmityl–enzyme, which is hydrolyzed to yield the **final products, palmitate and fatty acid synthase.**

E. **Palmitate** serves as the precursor for longer and unsaturated fatty acids. Chain-lengthening and desaturating systems allow synthesis of a variety of polyunsaturated fatty acids.

  1. Chain-lengthening systems are present in the mitochondria and the endoplasmic reticulum.

$$C_{16} \rightarrow C_{18} \rightarrow C_{20}$$

  2. A desaturating system is also present in the endoplasmic reticulum.

$$NADPH + H^+ + O_2 \qquad NADP^+ + H_2O$$

$$R{-}CH_2{-}CH_2{-}(CH_2)_7{-}COOH \longrightarrow R{-}\overset{10}{C}H{=}\overset{9}{C}H{-}(CH_2)_7{-}COOH$$

This desaturating system can insert double bonds no further than nine carbons from the carboxylic acid group.

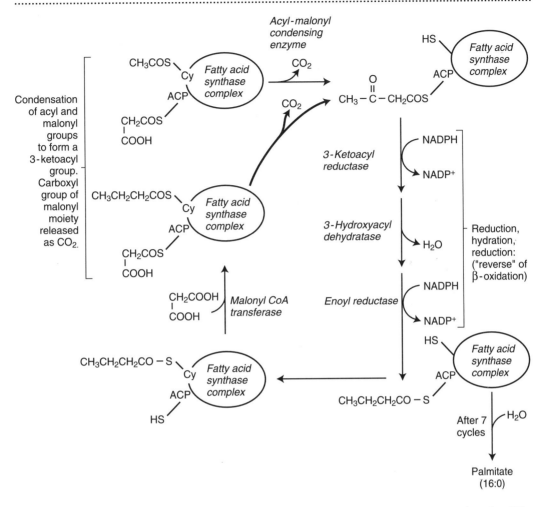

Figure 6-8. The reactions of fatty acid synthesis. ACP = acyl carrier protein; Cy = cysteinyl residue; HS = sulfhydryl group.

3. The limitations of the desaturating system impose a dietary requirement for **essential fatty acids** (those with double bonds > 10 carbons from the carboxyl end). **Lineoleic acid, linolenic acids,** which are essential fatty acids, fulfill this need.

4. Linoleic acid serves as a precursor for **arachidonic acid,** which is the beginning of the arachidonate cascade that synthesizes **prostaglandins, thromboxanes,** and **eicosanoids.**

VI. GLYCEROLIPID SYNTHESIS. This process is carried out by the liver, adipose tissue, and the intestine (Figure 6-9).

A. The pathways begin with **glycerol 3-phosphate,** which is mainly produced by reducing dihydroxyacetone phosphate with NADH.

B. Successive transfers of acyl groups from acyl CoA to carbons 1 and 2 of glycerol 3-phosphate produce **phosphatidate,** which can then be converted to a variety of lipids.

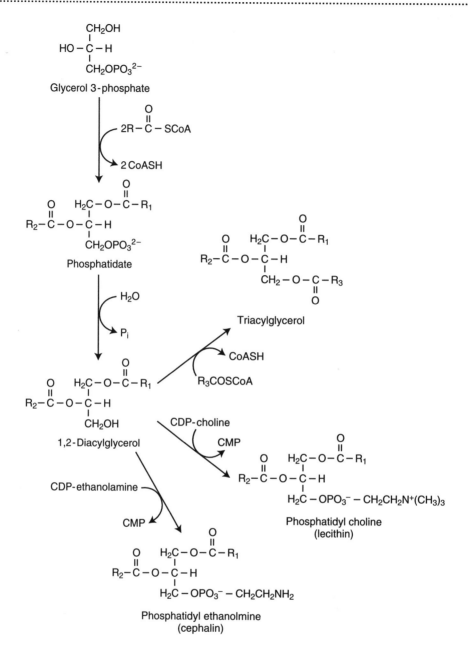

Figure 6-9. Synthesis of the major phospholipids. *CDP* = cytidine diphosphate; *CMP* = cytidine monophosphate.

1. **Triacylglycerol,** which results from the transfer of an acyl group from acyl CoA.

2. **Phosphatidyl choline** and **phosphatidyl ethanolamine,** which result from transfer of the base from its cytidine diphosphate (CDP) derivative.

3. **Phosphatidylserine,** which results from the exchange of serine for choline.

4. **Phosphatidylinositol,** which results from reaction of CDP-diacylglycerol with inositol.

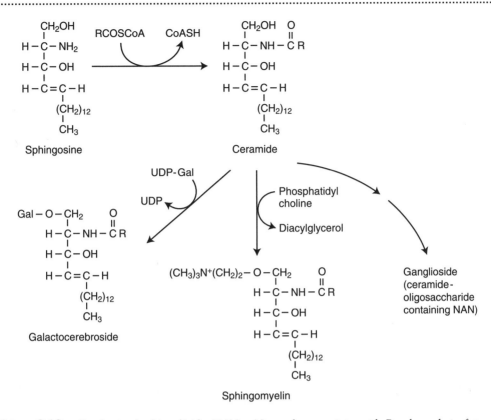

Figure 6-10. Synthesis of sphingolipids. *NAN* = N-acetyl neuraminic acid; *R* = long-chain fatty acid; *UDP-Gal* = UDP-galactose.

VII. SPHINGOLIPID SYNTHESIS (Figure 6-10)

A. The synthesis of sphingolipids, which do not contain glycerol, begins with palmityl CoA and serine. These substances are used to make **dihydrosphingosine** and **sphingosine.**

B. When sphingosine is acylated on the C_2-NH_2, **ceramide** is produced. Additional groups may be added to the C_1-OH of ceramides.

VIII. CHOLESTEROL SYNTHESIS

A. Cholesterol is synthesized by the liver and intestinal mucosa from **acetyl CoA** in a multistep process (Figure 6-11).

B. The **key intermediate** in cholesterol synthesis is **HMG CoA.**

  1. The regulatory enzyme is **HMG CoA reductase,** the reductant is NADPH, and the product is mevalonic acid.

  2. Increasing amounts of intracellular cholesterol lead to inhibition of HMG CoA reductase and accelerated degradation of the enzyme.

C. Mevalonic acid is the precursor of a number of natural products called **terpenes,** which include vitamin A, vitamin K, coenzyme Q, and natural rubber.

D. Cholesterol is also converted to the **steroid hormones** in the **adrenal cortex, ovary, placenta,** and **testes.**

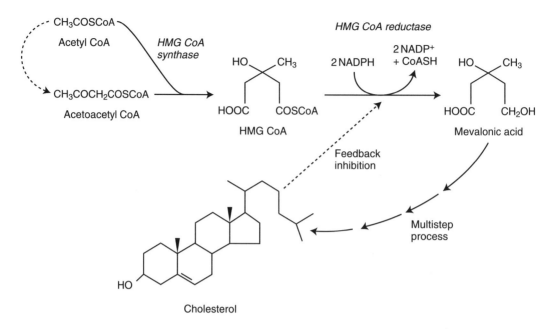

Figure 6-11. Sketch of cholesterol synthesis. *Italicized terms* = enzyme names.

E. The majority of cholesterol is oxidized to bile acids in the liver. 7-Dehydrocholesterol is the starting point for synthesis of **vitamin D.**

IX. CLINICAL RELEVANCE

A. **Lipid malabsorption** leading to excessive fat in the feces (**steatorrhea**) occurs for a variety of reasons.

    1. **Bile duct obstruction.** About 50% of the dietary fat appears in the stools as **soaps** (metal salts of LCFAs). The absence of bile pigments leads to **clay-colored stools,** and **deficiency** of the **ADEK vitamins** may result.

    2. **Pancreatic duct obstruction.** The stool contains **undigested fat.** Absorption of ADEK vitamins is not sufficiently impaired to lead to deficiency symptoms.

    3. **Diseases of the small intestine** (e.g., celiac disease, abetalipoproteinemia, nontropical sprue, inflammatory bowel disease) may impair lipid absorption.

B. Hyperlipidemias

    1. Defective LDL receptors lead to **familial hypercholesterolemia.**

        a. There is severe atherosclerosis and early death from coronary artery disease.

        b. Treatment with **HMG CoA reductase inhibitors** such as lovastatin or pravastatin can lower the blood cholesterol.

    2. **Hypertriglyceridemia** can result from either overproduction of VLDL or defective lipolysis of VLDL triglycerides. Cholesterol levels may be moderately increased.

    3. In **mixed hyperlipidemias** both serum cholesterol and serum triglycerides are elevated.

        a. There is both overproduction of VLDL and defective lipolysis of triglyceride-rich lipoproteins (VLDL and chylomicrons).

        b. There is danger of acute pancreatitis.

C. Clinical expression of disruptions in fatty acid oxidation

   1. Inherited defects in the carnitine transport system, which have widely varying symptoms.

      a. Hypoglycemia and some degree of muscle damage and muscle pain are usually present.

      b. Muscle wasting with accumulation of fat in muscle may occur in severe forms.

      c. Feeding fat with medium-chain triacylglycerols (e.g., butterfat) is helpful in some cases, because MCFAs can bypass the carnitine transport system.

   2. Inherited deficiencies in the acyl CoA dehydrogenases, the most common being medium-chain (C_6 to C_{12}) acyl CoA dehydrogenase deficiency.

      a. Hypoketotic hypoglycemia and dicarboxylic aciduria occur, with vomiting, lethargy, and coma.

      b. This is believed to account for the condition called "Reye-like syndrome."

D. Sphingolipidoses. Sphingolipids are normally degraded within the lysosomes of phagocytic cells. A number of **sphingolipid storage diseases** may occur (Table 6-1) as a result of deficiency of one of the lysosomal enzymes.

Table 6-1.
Sphingolipid Storage Disorders

Disorder	Accumulated Substance	Clinical Manifestations
Tay-Sachs disease	Ganglioside GM_2	Mental retardation, blindness, cherry-red spot on macula, death by third year
Gaucher's disease	Glucocerebroside	Liver and spleen enlargement, bone erosion, mental retardation (sometimes)
Fabry's disease	Ceramide trihexoside	Skin rash, kidney failure, lower extremity pain
Niemann-Pick disease	Sphingomyelin	Liver and spleen enlargement, mental retardation
Globoid cell leukodystrophy (Krabbe's disease)	Galactocerebroside	Mental retardation, myelin absent
Metachromatic leukodystrophy	Sulfatide	Mental retardation, metachromasia; nerves stain yellowish brown with crystal violet dye
Generalized gangliosidosis	Ganglioside GM_1	Mental retardation, liver enlargement, skeletal abnormalities
Sandhoff's disease	Ganglioside GM_2, globoside	Same as Tay-Sachs disease, but more rapid course
Fucosidosis	Pentahexosylfucoglycolipid	Cerebral degeneration, spasticity, thick skin

7

Amino Acid Metabolism

I. FUNCTIONS OF AMINO ACIDS

A. The **synthesis of new proteins** requires amino acids. The primary source of amino acids is dietary protein. Breakdown of tissue proteins also provides amino acids.

B. Amino acids provide nitrogen-containing substrates for the **biosynthesis** of:
  1. Nonessential amino acids
  2. Purines and pyrimidines
  3. Porphyrins
  4. Neurotransmitters and hormones

C. The **carbon skeletons** of the surplus amino acids not needed for synthetic pathways serve as **fuel**. They may be:
  1. Oxidized in the tricarboxylic acid (TCA) cycle to produce energy.
  2. Used as substrates for gluconeogenesis.
  3. Used as substrates for fatty acid synthesis.

II. REMOVAL OF AMINO ACID NITROGEN

A. **Deamination,** the first step in metabolizing surplus amino acids, **yields an α-keto acid and an ammonium ion (NH_4^+).**

B. **Transdeamination** effects deamination through the sequential actions of the enzymes transaminase **(aminotransferase)** and **glutamate dehydrogenase** (Figure 7-1).

C. The appearance of aspartate aminotransferase (AST) or alanine aminotransferase (ALT) in the blood is an indication of tissue damage, especially cardiac muscle (AST) and liver (AST and ALT).

III. UREA CYCLE AND DETOXIFICATION OF NH_4^+

A. NH_4^+ is toxic to the human body, particularly the central nervous system (CNS).

B. **Conversion of NH_4^+ to urea** occurs in the **liver** via the **urea cycle.** Urea is excreted in the **urine** (Figure 7-2).

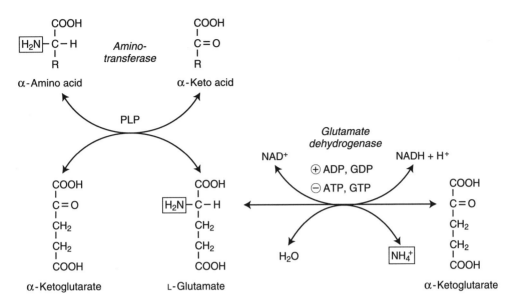

Figure 7-1. Deamination of an amino acid by the sequential action of an aminotransferase and glutamate dehydrogenase. α-Ketoglutarate and glutamate are a corresponding α-keto acid-amino acid pair. *PLP* = pyridoxal phosphate; ⊕ = activation; ⊖ = inhibition; *italicized terms* = enzyme names.

C. In **peripheral tissues,** detoxification of NH_4^+, which is ultimately converted to urea in the liver, occurs by different mechanisms.

    1. In most tissues, the enzyme **glutamine synthetase** incorporates NH_4^+ into **glutamate** to form **glutamine,** which is carried by the circulation to the liver. There the enzyme **glutaminase** hydrolyzes glutamine back to NH_4^+ and **glutamate.**

    2. In skeletal muscle, sequential action of the enzymes **glutamate dehydrogenase** and glutamate–pyruvate **aminotransferase** can lead to the incorporation of NH_4^+ into **alanine.** The alanine is carried to the liver, where **transdeamination** results in the conversion of the alanine back to pyruvate and NH_4^+.

D. Hyperammonemia

    1. This condition may be caused by insufficient removal of NH_4^+, resulting from disorders that involve one of the enzymes in the urea cycle.

    2. Blood ammonia concentrations above the normal range (30–60 μM) may cause coma due to ammonia intoxication.

    3. Ammonia intoxication can lead to mental retardation, seizure, coma, and death.

    4. Enzyme defects

        a. When the activity of the enzyme carbamoyl phosphate synthetase or ornithine–carbamoyl transferase is low, ammonia concentrations in the blood and urine rise, and ammonia intoxication can occur.

        b. When any of the enzymes argininosuccinate synthetase, argininosuccinase, or arginase is defective, blood levels of the metabolite immediately preceding the defect increase. Ammonia levels may also rise.

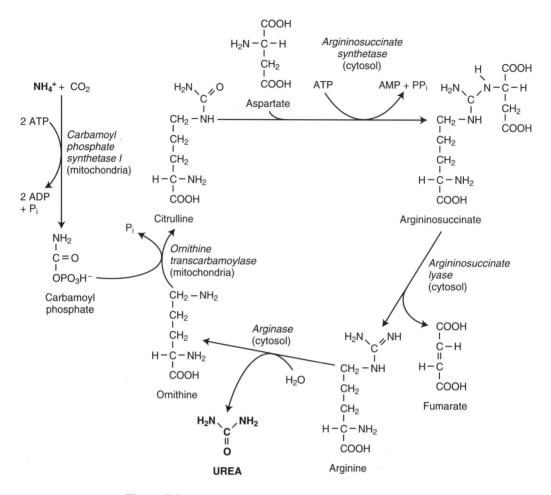

Figure 7-2. The urea cycle. *Italicized terms* = enzyme names.

   5. Treatment consists of restriction of dietary protein, intake of mixtures of keto acids that correspond to essential amino acids, and feeding benzoate and phenylacetate to provide an alternate pathway for ammonia excretion.

IV. CARBON SKELETONS OF AMINO ACIDS. The amino acids can be grouped into families based on the point where their carbon skeletons, the structural portions that remain after deamination, enter the TCA cycle (Figure 7-3 and Table 7-1).

   A. The amino acid carbon skeletons undergo a series of reactions whose products may be **glucogenic, ketogenic, or both.**

   B. Acetyl CoA family (also called the **ketogenic** amino acid family) [isoleucine, leucine, lysine, phenylalanine, tryptophan, and tyrosine]

      1. Acetyl CoA is the starting point for ketogenesis but cannot be used for net gluconeogenesis. **Leucine** and **lysine** are **only ketogenic** amino acids. The other four amino acids that form acetyl CoA are **both ketogenic and glucogenic.**

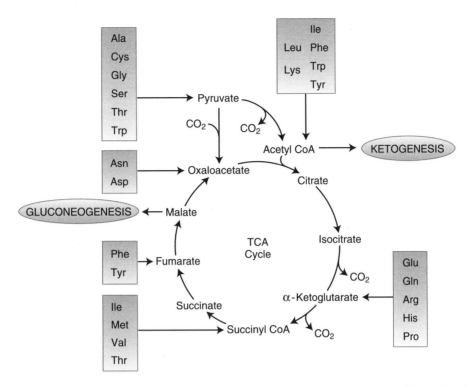

Figure 7-3.   Diagram showing where the amino acids enter the tricarboxylic acid (TCA) cycle.

   2. The first step in **phenylalanine** metabolism is conversion to tyrosine by the enzyme phenylalanine hydroxylase. **Tyrosine** is the starting compound for synthesizing some important products (Figure 7-4):

   a. **Epinephrine** and **norepinephrine**, which are **catecholamine hormones** that are secreted by the **adrenal medulla**

   b. **Triiodothyronine** and **thyroxine,** which are hormones that are secreted by the **thyroid** gland

   c. **Dopamine** and **norepinephrine,** which are **catecholamine neurotransmitters**

   d. **Melanin,** the pigment of skin and hair

C. α-Ketoglutarate family (arginine, histidine, glutamate, glutamine, and proline)

   1. **Histidine** is the precursor of **histamine,** a substance released by mast cells during **inflammation.**

   2. **Glutamate** is an **excitatory neurotransmitter.** In addition, it can be converted to the inhibitory **neurotransmitter** γ-aminobutyric acid **(GABA).**

D. Succinyl CoA family (isoleucine, methionine, and valine)

   1. The sulfur atom of **methionine** can be used in cysteine synthesis.

   2. The methyl group of **methionine** can participate in methylation reactions as *S*-adenosylmethionine **(SAM).**

E. **Fumarate** family (phenylalanine and tyrosine)

Table 7-1.
Amino Acids Classified By Point of Entrance
into the Tricarboxylic Acid (TCA) Cycle

TCA Cycle Substrate	Amino Acids
Acetyl CoA	Isoleucine* Leucine* Lysine* Phenylalanine* Tryptophan* Tyrosine
α-Ketoglutarate	Arginine Histidine* Glutamate Glutamine Proline
Succinyl CoA	Isoleucine* Methionine* Valine* Threonine*
Fumarate	Phenylalanine* Tyrosine
Oxaloacetate	Asparagine Aspartate
Pyruvate	Alanine Cysteine Glycine Serine Threonine* Tryptophan*

CoA = coenzyme A
* These are essential amino acids, which cannot be synthesized in the body, so
they must come from diet.

F. Oxaloacetate family (asparagine and aspartate)

G. Pyruvate family (alanine, cysteine, glycine, serine, threonine, and tryptophan)

   1. The **sulfhydryl** groups of **cysteine** residues produce **sulfate** ions.

   2. Glycine and **serine** can furnish **one-carbon** groups for the tetrahydrofolate one-carbon pool.

   3. Tryptophan is the precursor of the neurotransmitter **serotonin.**

V. CLINICAL RELEVANCE: INHERITED (INBORN) ERRORS OF AMINO ACID METABOLISM

   A. Phenylketonuria (PKU)

   1. Phenylalanine accumulates in the blood (hyperphenylalaninemia).

      a. Phenylalanine builds up to **toxic concentrations** in body fluids, resulting in CNS damage with mental retardation.

      b. Elevated phenylalanine inhibits **melanin synthesis,** leading to hypopigmentation.

   2. This condition results from a deficiency of **phenylalanine hydroxylase** or **dihydropteridine reductase** (see Figure 7-4).

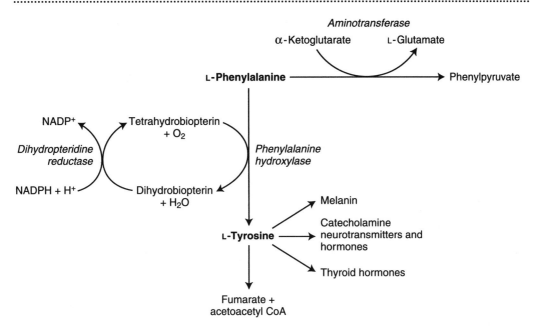

Figure 7-4. Catabolic pathways for phenylalanine and tyrosine. *Italicized terms* = enzyme names.

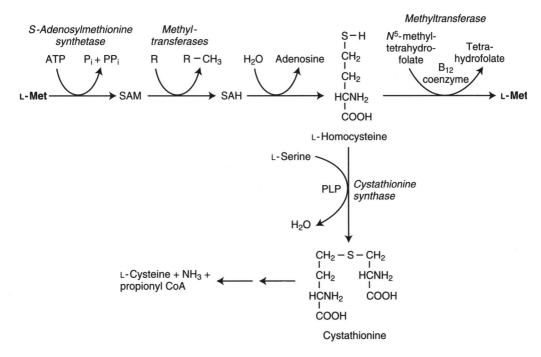

Figure 7-5. Metabolism of methionine. *L-Met* = L-Methionine; *SAH* = S-adenosyl homocysteine; *SAM* = S-adenosylmethionine; *PLP* = pyridoxal phosphate; *italicized terms* = enzyme names.

3. An alternative pathway for phenylalanine breakdown produces **phenylketones** (phenylpyruvic, phenyllactic, and phenylacetic acids), which spill into the urine.

4. In affected individuals, **tyrosine** is an **essential dietary amino acid.**

5. Treatment consists of restricting dietary protein (phenylalanine).

B. Albinism

1. Tyrosinase, the first enzyme on the pathway to melanin, is absent.

2. Albinos have little or no **melanin** (skin pigment). They **sunburn** easily, and are:
  a. Particularly susceptible to **skin carcinoma.**
  b. Photophobic because of lack of pigment in the iris of the eye.

C. Homocystinuria

1. In this disorder, **homocysteine,** which accumulates in blood and body fluids, appears in the urine.

2. Homocystinuria may result from several defects (Figure 7-5).
  a. Cystathionine synthase deficiency
  b. Reduced affinity of cystathionine synthase for its coenzyme, **pyridoxal phosphate (PLP)** [This form may respond to megadoses of pyridoxine (vitamin B_6).]
  c. N^5-Methyl tetrahydrofolate homocysteine methyltransferase deficiency

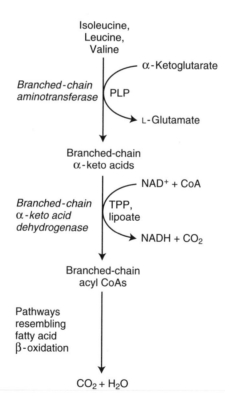

Figure 7-6.   Initial steps in the breakdown of the branched-chain amino acids. *PLP* = pyridoxal phosphate; *TPP* = thiamine pyrophosphate.

    d. Vitamin B$_{12}$ coenzyme (methylcobalamin) deficiency [This form may respond to vitamin B$_{12}$ supplements.]

3. Pathologic changes
    a. Dislocation of the optic lens
    b. Mental retardation
    c. Osteoporosis and other skeletal abnormalities
    d. Atherosclerosis and thromboembolism

4. Patients who are unresponsive to vitamin therapy may be treated with synthetic diets low in methionine, and by administering betaine (*N,N,N*-trimethylglycine) as an alternative methyl group donor.

D. Maple-syrup urine disease

1. In this disorder, the branched-chain keto acids derived from isoleucine, leucine, and valine appear in the urine, giving it a maple syrup-like odor.

2. This condition results from a deficiency in the branched-chain 2-keto acid decarboxylase (Figure 7-6).

3. The elevated keto acids cause severe brain damage, with death in the first year of life.

4. Treatment. A few cases respond to megadoses of thiamine (vitamin B$_1$). Otherwise, synthetic diets low in branched-chain amino acids are given.

E. Histidinemia

1. This disorder is characterized by elevated histidine in the blood plasma and excessive histidine metabolites in the urine.

2. The enzyme **histidine-α-deaminase,** the first enzyme in histidine catabolism, is deficient.

3. Mental retardation and speech defects may occur, but are rare.

4. Treatment is not usually indicated.

8

Nucleotide Metabolism

I. NUCLEOTIDE STRUCTURE

A. Nucleotides contain **three units** (Figure 8-1).

   1. Sugar (ribose or deoxyribose)

   2. Base

   a. Purines: adenine (A); guanine (G)
   b. Pyrimidines: cytosine (C); thymine (T); uracil (U)

   3. Phosphate group (at least one)

B. A **nucleoside** is a sugar with a base in a glycosidic linkage to C1, and a **nucleotide** is a nucleoside with one or more phosphate groups in an ester linkage to C5 (i.e., a nucleotide is a phosphorylated nucleoside).

II. NUCLEOTIDE FUNCTION

A. Substrates for DNA synthesis (replication): dATP, dGTP, dTTP, dCTP

B. Substrates for RNA synthesis (transcription): ATP, GTP, UTP, CTP

C. Carriers of high-energy groups

   1. Phosphoryl groups: ATP, UTP, GTP

   2. Sugar moieties: UDP glucose, CDP choline

   3. Acyl groups: acetyl CoA, acyl CoA

   4. Methyl groups: S-adenosylmethionine

D. Components of coenzymes: NAD, NADP, FAD, CoA

E. Regulatory molecules: cyclic AMP, cyclic GMP

III. PURINE NUCLEOTIDE SYNTHESIS

A. Origin of the atoms in the purine ring (Figure 8-2)

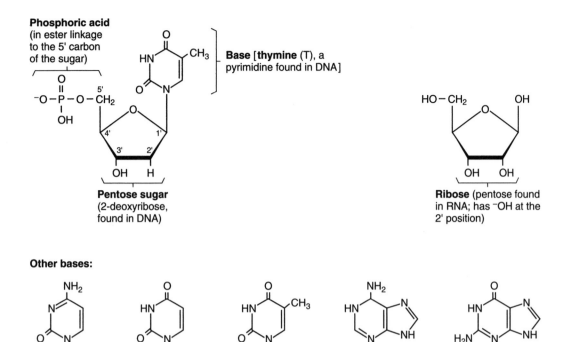

Figure 8-1. The general structure of nucleotides.

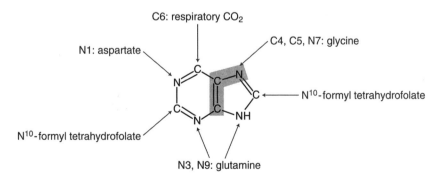

Figure 8-2. Origin of the atoms in the purine ring.

B. De novo purine nucleotide synthesis (Figure 8-3)

1. **Synthesis of 5'-phosphoribosyl-1-pyrophosphate (PRPP)** occurs at the beginning of the process. Inosine monophosphate (IMP), AMP, and GMP inhibit the enzyme PRPP synthetase.

2. The **committed step** involves the conversion of **PRPP to 5'-phosphoribosyl-1-amine.** PRPP activates the enzyme glutamine PRPP amidotransferase, and the end products of the pathway inhibit the enzyme. These end products are:

a. **IMP,** which is formed on the amino group of phosphoribosylamine by a nine-reaction sequence.

b. **GMP,** which is formed by the addition of an amino group to C2 of IMP.

c. **AMP,** which is formed by substitution of an amino group for the oxygen at C6.

C. Regulation of purine synthesis

1. Regulation occurs in the final branches of the de novo pathway to provide a **steady supply** of purine nucleotides.

a. GMP and AMP both inhibit the first step in their own synthesis from IMP.

b. GTP is a substrate in AMP synthesis, and ATP is a substrate in GMP synthesis. This is known as the **reciprocal substrate effect.** It balances the supply of adenine and guanine ribonucleotides.

2. Interconversion among purine nucleotides ensures control of the levels of adenine and guanine nucleotides.

a. AMP deaminase converts AMP back to IMP.

b. GMP reductase converts GMP back to IMP.

c. IMP is the starting point for synthesis of AMP and GMP.

D. Purine nucleotides can also by synthesized by **salvage** of preformed purine bases. This process involves two enzymes:

1. Hypoxanthine-guanine phosphoribosyltransferase (HGPRT) [Figure 8-4]. IMP and GMP are competitive inhibitors of HGPRT.

2. Adenine phosphoribosyl transferase. AMP inhibits this enzyme.

IV. PYRIMIDINE NUCLEOTIDE SYNTHESIS

A. Origin of atoms in the pyrimidine ring (Figure 8-5)

B. De novo pyrimidine synthesis (Figure 8-6)

1. **Synthesis of carbamoyl phosphate (CAP)** occurs at the beginning of the process, using CO_2 and glutamine, with the cytosolic enzyme carbamoyl phosphate synthetase II, which differs from the mitochondrial enzyme in the urea cycle.

2. The **synthesis of dihydroorotic acid,** a pyrimidine, is a two-step process.

a. The committed step is the addition of aspartate to CAP, which is catalyzed by the enzyme aspartate transcarbamoylase, to form carbamoyl aspartate.

b. Ring closure via loss of H_2O, which is catalyzed by the enzyme dihydroorotase, produces **dihydroorotic acid,** a pyrimidine.

3. In mammals, these first three steps of pyrimidine biosynthesis occur on a **single multifunctional enzyme** called **CAD,** which stands for the names of the enzymes (i.e., carbamoyl phosphate synthetase, aspartate transcarbamoylase, dihydroorotase).

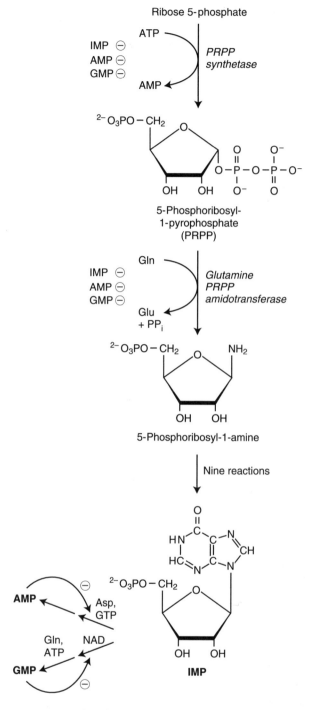

Figure 8-3.   De novo purine nucleotide synthesis. The end products IMP, GMP, and AMP inhibit the enzyme glutamine PRPP amidotransferase. ⊖ = inhibitor; *italicized terms* = enzyme names.

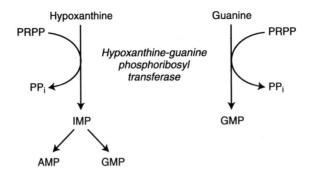

Figure 8-4. Purine nucleotide salvage by hypoxanthine-guanine phosphoribosyl transferase. *Italicized term* = enzyme name; *PRPP* = 5-phosphoribosyl-1-pyrophosphate.

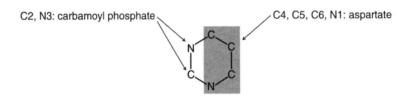

Figure 8-5. Origin of the atoms in the pyrimidine ring.

4. **Dihyroorotate forms UMP,** a pyrimidine nucleotide.
   a. Addition of a ribose-phosphate moiety from PRPP by orotate phosphoribosyltransferase yields **orotidylate (OMP).**
   b. Decarboxylation of **OMP** forms **uridylate (UMP).**
   c. These two steps occur on a single protein, and a defect in this protein leads to **orotic aciduria.**

5. **Synthesis of the remaining pyrimidine ribonucleotides involves UMP.**
   a. Phosphorylation of UMP results in the formation of UDP and UTP, at the expense of ATP.
   b. The addition of an amino group from glutamine to UTP yields CTP. Low concentrations of GTP activate the enzyme.

C. **Regulation of pyrimidine synthesis** occurs at several levels:
   1. UTP inhibits carbamoyl phosphate synthetase II, and ATP and PRPP activate this enzyme.
   2. UMP and CMP (to a lesser extent) inhibit OMP decarboxylase.
   3. CTP itself inhibits CTP synthetase.

D. **Salvage** of pyrimidines is accomplished by the enzyme **pyrimidine phosphoribosyl transferase,** which can use orotic acid, uracil or thymine, but not cytosine.

E. With ATP as the source of high-energy phosphate (~P), several enzymes provide a supply of nucleoside di- and triphosphates.
   1. Adenylate kinase catalyzes **interconversion among AMP, ADP, and ATP.**

$$\text{AMP} + \text{ATP} \rightleftharpoons 2\text{ADP} \qquad (\text{K}_{eq} \cong 1)$$

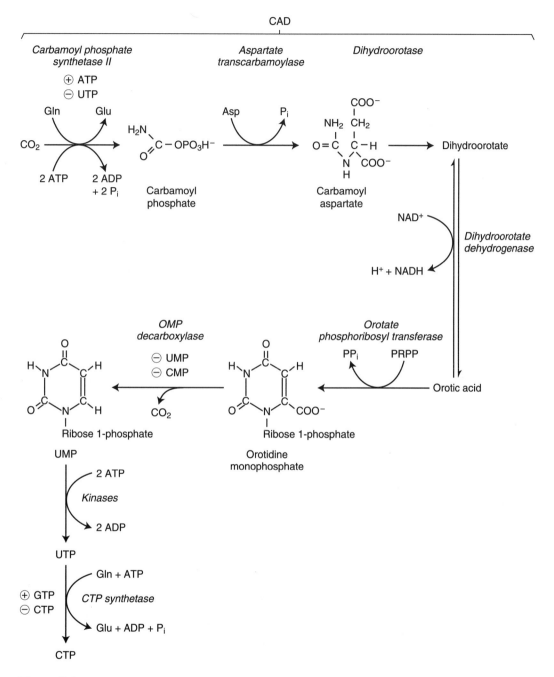

Figure 8-6. De novo pyrimidine synthesis. ⊕ = activator; ⊖ = inhibitor; *italicized terms* = enzyme names.

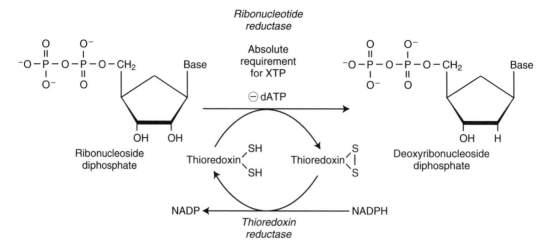

Figure 8-7. Deoxyribonucleotide synthesis. $\ominus$ = inhibitor; *italicized terms* = enzyme names.

2. Nucleoside monophosphate kinases provide the nucleoside diphosphates. For example:

$$UMP + ATP \rightleftharpoons UDP + ADP$$

3. Nucleoside diphosphate kinase, an enzyme with broad specificity, provides the nucleoside triphosphates. For example,

$$XDP + ATP \rightleftharpoons XTP + ADP$$

where X is a ribonucleoside or deoxyribonucleoside

V. DEOXYRIBONUCLEOTIDE SYNTHESIS

A. Formation of **deoxyribonucleotides**, which are required for DNA synthesis, involves the reduction of the sugar moiety of **ribonucleoside diphosphates**.

1. The complex enzyme **ribonucleotide reductase** leads to the reduction of ADP, GDP, CDP, or UDP to the deoxyribonucleotides (Figure 8-7).

a. The reducing power of this enzyme derives from two sulfhydryl groups on the small protein **thioredoxin**.

b. Using NADPH + H+, the enzyme **thioredoxin reductase** converts oxidized thioredoxin back to the reduced form.

2. **Strict regulation of ribonucleotide reductase** controls the overall supply of deoxyribonucleotides.

a. The reduction reaction proceeds only in the presence of a nucleoside triphosphate.

b. dATP is an allosteric inhibitor, so rising dATP levels will slow down the formation of all the deoxyribonucleotides.

c. The other deoxynucleoside triphosphates interact with allosteric sites to alter the substrate specificity.

B. The enzyme **thymidylate synthase catalyzes the formation of deoxythymidylate (dTMP)** from dUMP (Figure 8-8).

1. Transfer of a one-carbon unit from N^5,N^{10}-methylene tetrahydrofolate (FH$_4$) to C5 of the uracil ring occurs.

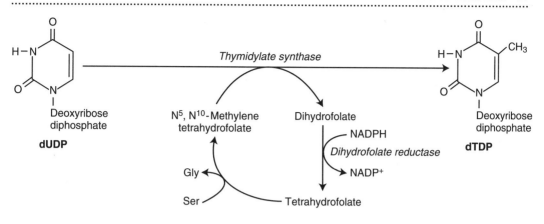

Figure 8-8. Thymidylate synthesis.

2. Simultaneously, the methylene group is reduced to a methyl group, with FH_4 serving as the reducing agent. The FH_4 is oxidized to dihydrofolate.

3. The coenzyme must be regenerated.

   a. Dihydrofolate is reduced by the enzyme **dihydrofolate reductase**, with NADPH as the reducing cofactor.

   b. Tetrahydrofolate is methylated at the expense of serine.

VI. NUCLEOTIDE DEGRADATION

A. Purine degradation. One of the products of purine nucleotide degradation is **uric acid,** which is excreted in the urine (Figure 8-9).

1. The sequential actions of two groups of enzymes, nucleases and nucleotidases, lead to the hydrolysis of nucleic acids to nucleosides.

2. The enzyme adenosine deaminase converts adenosine and deoxyadenosine to inosine or deoxyinosine.

3. Purine nucleoside phosphorylase splits inosine and guanosine to ribose 1-phosphate and the free bases hypoxanthine and guanine.

4. Guanine is deaminated to xanthine.

5. Hypoxanthine and xanthine are oxidized to uric acid by the enzyme xanthine oxidase.

B. Pyrimidine degradation. The products of degradation are β-amino acids, CO_2, and NH_4^+.

1. Surplus nucleotides are degraded to the free bases uracil or thymine.

2. A three-enzyme reaction sequence consisting of reduction, ring opening, and deamination-decarboxylation converts uracil to CO_2, NH_4+, and β-alanine.

3. The same enzymes convert thymine to CO_2, NH_4^+, and β-aminoisobutyrate. **Urinary β-aminoisobutyrate,** which originates exclusively from thymine degradation, is therefore an **indicator of DNA turnover.** It may be elevated during chemotherapy or radiation therapy.

VII. CLINICAL RELEVANCE

A. Disorders caused by deficiencies in enzymes involved in nucleotide metabolism

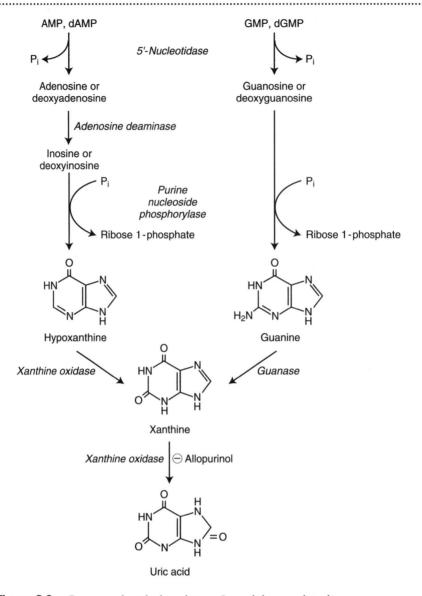

Figure 8-9. Purine nucleotide degradation. ⊖ = inhibitor; *italicized terms* = enzyme names.

1. **Hereditary orotic aciduria**

    a. **Enzyme:** orotate phosphoribosyl transferase and/or OMP decarboxylase

    b. **Characteristics:** retarded growth and severe anemia

    c. **Treatment:** feeding of synthetic cytidine or uridine supplies the pyrimidine nucleotides needed for RNA and DNA synthesis, restores normal growth, and reverses the anemia. UTP formed from these nucleosides acts as a feedback inhibitor of carbamoyl phosphate synthetase II, thus shutting down orotic acid synthesis.

2. **Purine nucleoside phosphorylase deficiency** leads to increased levels of purine nucleosides, with decreased uric acid formation. There is **impaired T-cell function**.

3. Severe combined immunodeficiency (SCID)

    a. Enzyme: adenosine deaminase

    b. Characteristics: T-cell and B-cell dysfunction with early death from overwhelming infection

    c. Treatment: SCID has been successfully treated by gene therapy.

4. Lesch-Nyhan syndrome

    a. Enzyme: HGPRTase (deficiency or absence of the salvage enzyme)

    b. Characteristics: excessive purine synthesis, hyperuricemia, and **severe neurologic problems,** which can include spasticity, mental retardation, and **self-mutilation**

        (1) No salvage of hypoxanthine and guanine occurs, so intracellular **IMP and GMP are decreased** and **the de novo pathway is not properly regulated.**

        (2) Intracellular **PRPP is increased, stimulating the de novo pathway.**

    c. Treatment: allopurinol decreases deposition of sodium urate crystals, but does not ameliorate the neurologic symptoms.

B. **Anticancer drugs that interfere with nucleotide metabolism**

    1. One of the hallmarks of cancer is rapidly dividing cells.

    2. Drugs that interfere with DNA synthesis inhibit (and sometimes stop) this rapid growth.

        a. Hydroxyurea inhibits **nucleoside diphosphate reductase,** the enzyme that converts ribonucleotides to deoxyribonucleotides.

        b. Aminopterin and **methotrexate** inhibit **dihydrofolate reductase,** the enzyme that converts dihydrofolate to tetrahydrofolate.

        c. Fluorodeoxyuridylate inhibits **thymidylate synthetase,** the enzyme that converts dUMP to dTMP.

C. Gout may result from a disorder in purine metabolism.

    1. Gout, a form of acute arthritis, is associated with **hyperuricemia** (elevated blood uric acid).

    2. Uric acid is not very soluble in body fluids. In hyperuricemia, sodium urate crystals are deposited in joints and soft tissues, causing the inflammation that characterizes arthritis. Crystals can also form in the kidney, leading to renal damage.

    3. In **primary gout, overproduction of purine nucleotides** occurs by de novo synthesis.

        a. Mutations occur in PRPP synthetase, with loss of feedback inhibition by purine nucleotides.

        b. A partial HGPRTase deficiency may develop, so that less PRPP is consumed by the salvage enzymes. Elevated PRPP activates PRPP amidotransferase.

    4. Increased cell death as a result of radiation therapy or cancer chemotherapy may elevate uric acid levels.

    5. Treatment. Primary gout is frequently treated with **allopurinol.**

        a. The enzyme **xanthine oxidase** catalyzes the oxidation of allopurinol to **alloxanthine,** which is a **potent inhibitor** of the enzyme.

        b. Uric acid levels fall, and hypoxanthine and xanthine levels rise.

        c. Hypoxanthine and xanthine are more soluble than uric acid, so they do not form crystal deposits.

9

Nutrition

I. ENERGY NEEDS

A. **Energy requirements** are expressed as either kilocalories (kcal) or joules (1 kcal = 4.184 kJ).

B. **Energy expenditure** (three components)

    1. The **basal energy expenditure (BEE),** which is also called the resting energy expenditure, is the energy used for metabolic processes while at rest. It represents more than 60% of the total energy expenditure. The BEE is related to the **lean body mass.**

    2. The **thermic effect of food,** the energy required for digesting and absorbing food, amounts to about 10% of the energy expenditure.

    3. The **activity-related expenditure,** which varies with the level of physical activity, represents 20% to 30% of the daily energy expenditure.

C. **Caloric requirements.** Table 9-1 gives the estimated **daily energy needs.**

Table 9-1.
Estimated Daily Energy Needs By Age

Age	kcal/lb DBW	kcal/kg DBW
Infants 0–12 months	~55	~120
Children 1–10 years (gradually decreases with age)	36–45	80–100
Young men 11–15 years	~30	~65
Young women 11–15 years	~17	~35
Young men 16–20 years, average activity	~18	~40
Young women ≥ 16 years	~15	~30
Adults	~13–15	~28–30

DBW = desirable body weight.

D. Caloric yield from foods

   1. Carbohydrates: 4 kcal/g

   2. Proteins: 4 kcal/g

   3. Fats: 9 kcal/g

   4. Alcohol: 7 kcal/g

II. MACRONUTRIENTS

A. Carbohydrates should comprise 50% to 60% of the caloric intake.

   1. Available carbohydrates

      a. Monosaccharides (e.g., glucose, fructose)

      b. Disaccharides (e.g., sucrose, lactose, maltose)

      c. Polysaccharides (e.g., starches, dextrins, glycogen)

   2. Unavailable carbohydrates, primarily **fiber,** are not digested and absorbed, but provide bulk and **assist elimination.**

      a. Insoluble fiber (e.g., cellulose, hemicellulose, and lignin) in unrefined cereals, bran, and some fruits and vegetables absorbs water, thus increasing stool bulk and shortening intestinal transit. (Lignin binds cholesterol and carcinogens.)

      b. Soluble fiber (e.g., pectins from fruits, gums from dried beans and oats) slows the rate of gastric emptying, decreases the rate of sugar uptake, and lowers serum cholesterol.

   3. Function. The tissues use carbohydrates (principally as glucose) for fuel after digestion and absorption have occurred.

   4. Inadequate carbohydrate intake (< 60 g/day) may lead to ketosis, excessive breakdown of tissue proteins (wasting), loss of cations (Na⁺), and dehydration.

   5. Excess carbohydrates are stored as glycogen and fat (triacylglycerol).

B. Fats should comprise no more than 30% of the caloric intake.

   1. The **essential fatty acids (EFAs)** are **linoleic acid** (9,12-octadecadienoic acid, an ω-6 fatty acid) and **linolenic acid** (9,12,15-octadecatrienoic acid, an ω-3 fatty acid).

   2. Functions

      a. EFAs provide precursors for synthesis of prostaglandins, prostacyclins, leukotrienes, and thromboxanes.

      b. Dietary fat serves as a carrier for the fat-soluble vitamins.

      c. Dietary fat slows gastric emptying, gives a sensation of fullness, and lends food a desirable texture and taste.

   3. EFA deficiency, which is rare in the United States, is primarily seen in low-birth-weight infants maintained on artificial formulas and adults on total parenteral nutrition. The characteristic symptom is a scaly dermatitis.

   4. Excess dietary fat is stored as triacylglycerol.

C. Protein should comprise 10% to 20% of the caloric intake.

   1. The nine **essential amino acids,** which cannot be synthesized in the body from nonprotein precursors, are histidine, isoleucine, leucine, lysine, methionine, phenylalanine, threonine, tryptophan, and valine.

2. Function. Proteins provide the amino acids for synthesizing proteins and nonprotein nitrogenous substances (see Chapters 7 and 10).

3. Nitrogen balance is the difference between nitrogen intake (primarily as protein) and nitrogen excretion (undigested protein in the feces; urea and ammonia in the urine). A healthy adult is in nitrogen balance, with excretion equal to intake.

   a. In **positive nitrogen balance,** intake exceeds excretion. This occurs when protein requirements increase (during pregnancy and lactation, growth, or recovery from surgery, trauma, or infection).

   b. In **negative nitrogen balance,** excretion exceeds intake. This occurs during metabolic stress, when dietary protein is too low, or when an **essential amino acid** is missing.

4. The **recommended adult protein intake** is 0.8 g/kg body weight/day, or about 60 g for a 75-kg (165-lb) person.

   a. This assumes easy digestion and absorption and composition of essential amino acids in a proportion similar to that of the human body. This is true for most animal proteins.

   b. Some vegetable proteins, which are more difficult to digest, are often low in one or more of the essential amino acids. Vegetarian diets may require higher protein intake, and they should include two or more different proteins to provide sufficient essential amino acids.

D. Clinical relevance: protein–energy malnutrition (PEM) syndromes

   1. Marasmus is caused by starvation, with insufficient intake of food, including both calories and protein. Signs and symptoms are numerous (Table 9-2).

Table 9-2.

Symptoms of Protein–Energy Malnutrition (PEM) Syndromes

Marasmus	Kwashiorkor
Depleted subcutaneous fat	Subcutaneous fat loss less extreme
	Pitting edema, usually in the feet and lower legs, but may affect most of the body
	Characteristic skin changes [dark patches that peel ("flaky paint" dermatosis)]
	Easily pluckable hair
Ketogenesis in the liver to provide fuel for brain and cardiac muscle	Enlarged liver due to fatty infiltration
Muscle wasting, as muscle proteins break down to provide amino acids for gluconeogenesis and hepatic protein synthesis	Muscle wasting less extreme
Frequent infections	Frequent infections
Low body temperature, except during infections	
Signs of micronutrient deficiencies	Other nutrient deficiencies
Slowed growth (< 60% of expected weight for age)	Growth failure (but > 60% of expected weight for age)
Death occurs when energy and protein reserves are exhausted	Poor appetite (anorexia)
	Frequent loose, watery stools containing undigested food particles
	Mental changes (apathetic and unsmiling, irritable when disturbed)

Table 9-3.

Health Risks Associated With Obesity

Body Mass Index	Status	Health Risk
20	Desirable	Acceptable
25	Overweight	Low
30	Obese	Moderate
35	Obese	High
40	Obese	Very high

2. Kwashiorkor is starvation with **edema.** This condition is often attributable to a diet more deficient in protein than total calories (see Table 9-2).

E. Obesity, which is an abnormally high percentage of body fat, is the most important nutritional problem in the United States, where 20% of adolescents and more than 30% of adults are overweight.

1. **Body fat** can be estimated by calculating the **body mass index (BMI)** [**Quetelet index**], which is the weight (kg) ÷ height (m) squared (BMI = kg/m^2).

2. The risk of poor health increases with increasing BMI (Table 9-3).

3. Diseases that may be associated with obesity

  a. High serum lipids, including cholesterol, and coronary artery disease
  b. Hypertension
  c. Non-insulin-dependent diabetes mellitus
  d. Cancer (breast and uterine)
  e. Gallstone formation
  f. Degenerative joint disease (osteoarthritis)
  g. Respiratory problems (inadequate ventilation, reduced functional lung volume)

III. MICRONUTRIENTS: THE FAT-SOLUBLE VITAMINS

A. Vitamin A

  1. Functions

   a. 11-*cis*-retinal is the prosthetic group of **rhodopsin,** the visual pigment in the rods and cones of the retina.
   b. β-carotene is an **antioxidant,** which protects against damage from free radicals.
   c. **Retinyl phosphate** serves as an acceptor/donor of mannose units in glycoprotein synthesis.
   d. **Retinol** and **retinoic acid** regulate tissue growth and differentiation.

  2. Sources

   a. Liver, egg yolks, and whole milk, which supply **retinol,** an active form of vitamin A
   b. Dark green and yellow vegetables, which supply **β-carotene,** a precursor of vitamin A
    (1) The body converts **β-carotene** to **retinol** and stores it in the liver.
    (2) Other active derivatives of **β-carotene** include **retinoic acid, retinyl phosphate,** and **11-*cis*-retinal.**

  3. **Recommended dietary allowance (RDA)** [adults]: 800–1000 μg retinol equivalents/day

Table 9-4.

Symptoms of Vitamin Deficiencies: The Fat-Soluble Vitamins

Vitamin	Deficiency-Associated Condition(s)
A	Night blindness
	Hyperkeratosis
	Anemia
	Xerophthalmia
	Low resistance to infection
	Increased risk for cancer
D	Rickets
	Osteomalacia
	Osteoporosis
E	Ataxia
	Myopathy
	Hemolytic anemia
	Retinal degeneration
K	Impaired blood clotting

4. Deficiency signs and symptoms (Table 9-4)

   a. Night blindness and xerophthalmia, or the progressive keratinization of the cornea, which is the leading cause of childhood blindness in developing nations
   b. Follicular hyperkeratosis, or rough, tough skin (i.e., like goosebumps)
   c. Anemia in the presence of adequate iron nutrition
   d. Decreased resistance to infection
   e. Increased susceptibility to cancer
   f. Impaired synthesis of serum retinol binding protein, with consequent inability to transport retinol to the tissues (apparent vitamin A deficiency; PEM or zinc deficiency)

5. Toxicity follows prolonged ingestion of 15,000 to 50,000 retinol equivalents/day.

   a. Signs and symptoms include bone pain, scaly dermatitis, enlarged liver and spleen, nausea, and diarrhea.
   b. Excess β-carotene is not toxic, because there is limited ability for liver conversion of the vitamin precursor to retinol.

6. Clinical usefulness of synthetic retinoids

   a. All *trans*-retinoic acids (tretinoin) and 13-*cis*-retinoic acid (isotretinoin), which are used in the **treatment of acne**
   b. Etretinate, a second-generation retinoid, which is used in the **treatment of psoriasis**

B. Vitamin D

   1. Functions, which include **regulation of calcium ion (Ca++) metabolism**

      a. Facilitation of absorption of dietary calcium by stimulation of synthesis of calcium-binding protein in the intestinal mucosa
      b. In combination with parathyroid hormone (PTH)
         (1) Promotion of bone demineralization by stimulation of osteoblast activity, thus releasing Ca++ into the blood
         (2) Stimulation of Ca++ reabsorption by the distal renal tubules, which also elevates blood Ca++

2. Sources

    a. Major source: the **skin,** where ultraviolet radiation converts 7-dehydrocholesterol to vitamin D_3 (cholecalciferol)

    b. Dietary sources of vitamin D_3: fish (marine), liver, and egg yolks

    c. Foods fortified with vitamin D_2 (ergocalciferol): dairy foods, margarine, and cereals

3. Activation

    a. Vitamin D is carried to the liver, where it is converted to 25-hydroxycholecalciferol [25(OH)D_3]. The kidney converts 25(OH)D_3 to the active form, **1,25(OH)$_2D_3$.**

    b. PTH, which is secreted in response to low serum calcium, stimulates the conversion to 1,25(OH)$_2D_3$.

4. Deficiency conditions (see Table 9-4)

    a. Rickets (young children): improperly mineralized, soft bones and stunted growth

    b. Osteomalacia (adults): demineralization of existing bones, with pathologic fractures

    c. Bone demineralization may also result from the conversion of vitamin D to inactive forms, which is stimulated by glucocorticoids.

5. RDA: 7.5 μg/day (in the absence of adequate sunlight)

6. Toxicity, which occurs with high doses (> 250 μg/day in adults, 25 μg/day in children), may lead to the following conditions:

    a. Hypercalcemia due to enhanced Ca^{++} absorption and bone resorption

    b. Metastatic **calcification** in soft tissue

    c. Bone demineralization

    d. Hypercalcuria, resulting in **kidney stones**

C. Vitamin E

    1. Functions include protection of membranes and proteins from free-radical damage. Vitamin E includes several isomers of tocopherol; the unit of potency is 1.0 mg RRR-α-tocopherol.

        a. The tocopherols function as free radical-trapping **antioxidants.**

        b. When tocopherol reacts with free radicals, it is converted to the tocopheroxyl radical. Vitamin C (ascorbic acid) reduces the tocopheroxyl radical and regenerates tocopherol.

    2. Sources: green leafy vegetables and seed grains

    3. RDA: 7 to 10 mg RRR-α-tocopherol equivalents

    4. Deficiency. Human vitamin E deficiency, which is secondary to impaired lipid absorption (see Table 9-4), may occur in diseases such as cystic fibrosis, celiac disease, chronic cholestasis, pancreatic insufficiency, and abetalipoproteinemia.

        a. Signs and symptoms include ataxia with impaired reflexes, myopathy with creatinuria, muscle weakness, hemolytic anemia, and retinal degeneration.

        b. Some signs and symptoms may be organ-specific, but they may also be nonspecific because they result from damage to cell membrane structures.

D. Vitamin K

    1. Function. Vitamin K is required for the **post-translational carboxylation of glutamyl residues** in a number of calcium-binding proteins, notably the **blood clotting factors VII, IX, and X.**

2. Sources

   a. Foods. Green vegetables are a good source of vitamin K (K_1, phylloquinone), and cereals, fruits, dairy products, and meats provide lesser amounts.
   b. Intestinal flora (microorganisms) also provide vitamin K (K_2, menaquinones)

3. RDA (adults): 65 to 80 μg (varies with varying production by the intestinal flora)

4. Deficiency. Vitamin K deficiency impairs blood clotting, with increased bruising and bleeding (see Table 9-4). **Causes** of deficiency include:

   a. Fat malabsorption
   b. Drugs that interfere with vitamin K metabolism
   c. Antibiotics that suppress bowel flora

5. Vitamin K in infants. Neonates are born with low stores of vitamin K.

   a. Vitamin K crosses the placental barrier poorly.
   b. Newborns are routinely given a single injection of vitamin K (0.5 to 1 mg), because they lack intestinal flora for synthesis of the vitamin.
   c. High doses can cause anemia, hyperbilirubinemia, and kernicterus (accumulation of bilirubin in the tissues).

IV. MICRONUTRIENTS: THE WATER-SOLUBLE VITAMINS

A. Thiamine (vitamin B_1)

   1. Functions. Thiamine pyrophosphate (TPP) is required for proper **nerve transmission.** TPP is the coenzyme for several **key enzymes.**

   a. Pyruvate and the α-ketoglutarate dehydrogenases (glycolysis and the citric acid cycle)
   b. Transketolase (the pentose phosphate pathway)
   c. Branched-chain keto-acid dehydrogenase (valine, leucine, and isoleucine metabolism)

   2. Sources: whole and enriched grains, meats, milk, and eggs

   3. RDA (adults): approximately 1 mg. The RDA, which is higher with a diet high in refined carbohydrates, decreases slightly with age.

   4. Deficiency (Table 9-5) leads to **beriberi,** which occurs in three stages:

   a. Early: loss of appetite, constipation and nausea, peripheral neuropathy, irritability, and fatigue
   b. Moderately severe: Wernicke-Korsakoff syndrome (seen in chronic alcoholics), which includes mental confusion, ataxia (unsteady gait, poor coordination), and ophthalmoplegia (loss of eye coordination)
   c. Severe
      (1) "**Dry beriberi**" includes all of the signs and symptoms in 4.a and 4.b plus more advanced neurologic symptoms, with atrophy and weakness of the muscles (e.g., foot drop, wrist drop).
      (2) "**Wet**" beriberi includes the symptoms of dry beriberi in combination with edema, high-output cardiac failure, and pulmonary congestion.

B. Riboflavin

   1. Function. Riboflavin is converted to the oxidation–reduction coenzymes flavin adenine dinucleotide (**FAD**) and flavin adenine mononucleotide (**FMN**).

   2. Sources: cereals, milk, meat, and eggs

Table 9-5.

Symptoms of Vitamin Deficiencies: The Water-Soluble Vitamins

Vitamin	Deficiency-Associated Condition(s)
Thiamine (vitamin B_1)	Wernicke-Korsakoff syndrome
	Beriberi
Riboflavin	Angular cheilitis
	Glossitis
	Scaly dermatitis
Niacin	Pellagra: dermatitis, diarrhea, dementia
Vitamin B_6	Irritability, depression
	Peripheral neuropathy, convulsions
	Eczema, dermatits
Pantothenic acid	Deficiency very rarely occurs
Biotin	Deficiency rarely occurs
	Symptoms include dermatitis, hair loss, etc.
Folic acid	Megaloblastic anemia
	Neural tube defects (maternal deficiency)
Vitamin B_{12}	Megaloblastic anemia
	Nervous system damage
Vitamin C (ascorbic acid)	Scurvy

3. RDA (adults): 1.0 to 1.7 mg

4. Deficiency signs and symptoms (see Table 9-5)

   a. Angular cheilitis, or inflammation and cracking at the corners of the lips

   b. Glossitis, or a red and swollen tongue

   c. Scaly dermatitis, particularly at the nasolabial folds and around the scrotum

C. Niacin (nicotinic acid) and niacinamide (nicotinamide) [vitamin B_6]

   1. Function. Niacin is converted to the oxidation–reduction coenzymes nicotinamide adenine dinucleotide (**NAD**) and nicotinamide adenine dinucleotide phosphate (**NADP**).

   2. Sources

      a. Whole and enriched cereals, milk, meats, and peanuts

      b. Synthesis from dietary tryptophan

   3. RDA: 12 to 20 mg of niacin or its equivalent (60 mg tryptophan = 1 mg niacin)

   4. Deficiency (see Table 9-5)

      a. Mild deficiency results in glossitis of the tongue.

      b. Severe deficiency leads to **pellagra**, characterized by the three Ds: **dermatitis, diarrhea,** and **dementia.**

   5. High doses (2 to 4 g/day) of nicotinic acid (not nicotinamide) result in vasodilation (**very rapid flushing**) and metabolic changes such as a decrease in blood cholesterol and low-density lipoproteins.

D. Vitamin B_6 (pyridoxine, pyridoxamine, and pyridoxal)

   1. Function. Pyridoxal phosphate is the coenzyme involved in **transamination** and other reactions of amino acid metabolism (see Chapter 7).

   2. Sources: whole grain cereals, nuts and seeds, vegetables, meats, eggs, and legumes

3. RDA (adults): 1.1 to 2.0 mg. The drugs isoniazid and penicillamine increase the requirement for vitamin B_6.

4. Deficiency (see Table 9-5)

   a. Mild: irritability, nervousness, and depression

   b. Severe: peripheral neuropathy and convulsions, with occasional sideroblastic anemia

   c. Other symptoms: eczema and seborrheic dermatitis around the ears, nose, and mouth; chapped lips; glossitis; and angular stomatitis

5. Clinical usefulness. High doses of vitamin B_6 are used to treat **homocystinuria** resulting from defective cystathionine β-synthase.

6. Prolonged consumption (> 500 mg/day) may lead to vitamin B_6 toxicity with sensory neuropathy.

E. Pantothenic acid

1. Function. Pantothenic acid is an essential component of **coenzyme A (CoA)** and the phosphopantetheine of **fatty acid synthase**.

2. Source: very widespread in food

3. RDA (none established)

4. Deficiency (very rare), with vague presentation that is of little concern to humans

F. Biotin

1. Function. Covalently linked biotin (biocytin) is the prosthetic group for **carboxylation enzymes** (e.g. pyruvate carboxylase, acetyl CoA carboxylase).

2. Sources

   a. Bacterial synthesis in the intestine

   b. Foods: organ meats, egg yolk, legumes, nuts, and chocolate

3. RDA: 100–200 μg/day. Biotin supplements are required during prolonged parenteral nutrition and in patients given long-term high-dose antibiotics.

4. Deficiency (rare) [see Table 9-5]

   a. Signs and symptoms include dermatitis, hair loss, atrophy of the lingual papillae, gray mucous membranes, muscle pain, paresthesia, hypercholesterolemia, and electrocardiographic abnormalities.

   b. Raw egg whites contain **avidin**, a protein that binds biotin in a nondigestible form; people who consume approximately 20 egg whites per day may develop biotin deficiency.

G. Folic acid (pteroylglutamic acid, folacin)

1. Function. Polyglutamate derivatives of **tetrahydrofolate** serve as coenzymes in one-carbon transfer reactions in purine and pyrimidine synthesis, thymidylate synthesis (see Chapter 8), conversion of homocysteine to methionine, and serine–glycine interconversion (see Chapter 7).

2. Sources: dark green leafy vegetables, meats, whole grains, and citrus fruits

3. RDA: 200 mg

4. Deficiency signs and symptoms (see Table 9-5)

   a. **Megaloblastic anemia,** similar to that of vitamin B_{12} deficiency, as a consequence of blocked DNA synthesis
   b. **Neural tube defects** as a result of maternal folate deficiency (in some cases)
   c. Elevated blood **homocysteine,** which is associated with **atherosclerotic heart disease,** with folate and vitamin B_6 deficiency (in some cases)
   d. Several **drugs** can lead to folate deficiency, including methotrexate (cancer chemotherapy), trimethoprim (antibacterial), pyrimethamine (antimalarial), and diphenylhydantoin and primidone (anticonvulsants).

H. Vitamin B_{12} (cobalamin)

   1. **Functions**

   a. Deoxyadenosyl cobalamin is the coenzyme for the conversion of methylmalonyl CoA to **succinyl CoA** (methylmalonyl CoA mutase) in the metabolism of propionyl CoA.
   b. Methylcobalamin is the coenzyme for methyl group transfer between tetrahydrofolate and methionine (homocysteine methyl transferase).

   2. **Sources**

   a. Meat, especially **liver;** fish; poultry; shellfish; eggs; and dairy products
   b. Vitamin B_{12} is not found in plant foods.

   3. **RDA:** 3 μg/day

   4. **Deficiency** signs and symptoms (see Table 9-5)

   a. **Megaloblastic anemia,** similar to that in folate deficiency
   b. Paresthesia (numbness and tingling of the extremities), with weakness and other neurologic changes
   c. Prolonged deficiency leads to **irreversible nervous system damage.**

   5. **Causes** of vitamin B_{12} deficiency

   a. **Intake of no animal products. Vegans** are at risk for vitamin B_{12} deficiency.
   b. **Impaired absorption** [from achlorhydria (insufficient gastric hydrochloric acid), decreased secretion of gastric intrinsic factor, impaired pancreatic function]

I. Vitamin C (ascorbic acid)

   1. **Functions**

   a. **Coenzyme for oxidation–reduction reactions,** including the post-translational hydroxylation of proline and lysine in the maturation of collagen, carnitine synthesis, tyrosine metabolism, and catecholamine neurotransmitter synthesis
   b. **Antioxidant**
   c. Facilitator of iron absorption

   2. **Sources:** fruits and vegetables

   3. **RDA:** 60 mg (increased in smokers)

   4. **Deficiency** signs and symptoms (see Table 9-5)

   a. **Mild deficiency:** capillary fragility with easy bruising and petechiae (pinpoint hemorrhages in the skin), as well as decreased immune function
   b. **Severe deficiency: scurvy,** with decreased wound healing, osteoporosis, hemorrhaging, and anemia; the teeth may fall out

V. MINERALS

A. **Calcium.** This mineral is the fifth most abundant element in the body and the most abundant cation.

   1. Functions

   a. Essential in the **formation of the bones and teeth** (99% of body calcium is in the bones)
   b. Essential for **normal nerve and muscle function.**
   c. Essential for **blood clotting.**

   2. **Sources:** dairy products (the most important source in the United States), as well as fortified fruit juices and cereals, fish with bones, collards, and turnip greens

   3. RDA: 1000 mg

   4. **Deficiency** signs and symptoms (Table 9-6)

   a. **Paresthesia** (tingling sensation), increased neuromuscular excitability, and muscle cramps. Severe hypocalcemia can lead to tetany.
   b. **Bone fractures,** bone pain, and loss of height
   c. **Osteomalacia** (as with vitamin D deficiency)

B. Iodine

   1. **Function:** incorporation into **thyroid hormones,** which is called **organification**

   2. **Sources: seafood and iodized salt** (iodine content of other foods varies depending on the soil)

   3. RDA: 120 to 150 mg

   4. **Deficiency** signs and symptoms (see Table 9-6)

   a. **Goiter** (enlarged thyroid gland)
   b. **Cretinism** (retarded growth and mental development)

   5. **Increased levels.** High iodine intake may cause goiter by blocking organification.

Table 9-6.
Symptoms of Mineral Deficiencies

Mineral	Deficiency-Associated Condition(s)
Calcium	Paresthesia
	Tetany
	Bone fractures, bone pain
	Osteomalacia (as in vitamin D deficiency)
Iodine	Goiter
	Cretinism
Iron	Anemia
	Fatigue, tachycardia, dyspnea
Magnesium	Neuromuscular excitability, paresthesia
	Depressed PTH release
Phosphorus (as phosphate)	Deficiency rarely occurs
Zinc	Growth retardation
	Dry, scaly skin
	Mental lethargy

PTH = parathyroid hormone.

C. Iron

    1. Functions (primarily due to the presence of iron in **heme molecules)**

        a. Oxygen transport (hemoglobin and myoglobin)

        b. Electron transport (cytochromes)

        c. Activation of oxygen (oxidases and oxygenases)

    2. Sources

        a. **Foods high in iron** include liver, heart, wheat germ, egg yolks, oysters, fruits, and some dried beans.

        b. **Foods with lesser amounts of iron** are muscle meats, fish, fowl, green vegetables, and cereals. Foods low in iron include dairy products and most nongreen vegetables.

    3. **RDA:** 10 mg (adult men); 18 mg (adult women)

    4. Absorption

        a. Heme iron is absorbed more efficiently (10% to 20%) than nonheme iron (< 10%).

        b. Ascorbic acid, reducing sugars, and meat enhance iron absorption.

        c. Antacids and certain plant food constituents (phytate, oxalate, fiber, tannin) may reduce iron absorption.

    5. Deficiency signs and symptoms (see Table 9-6)

        a. Hypochromic microcytic **anemia**

        b. **Fatigue,** pallor, tachycardia, dyspnea (shortness of breath) on exertion

        c. Burning sensation, with depapillation of the tongue

    6. Toxicity

        a. Excessive iron intake leads to **hemochromatosis**.

        b. Large doses of ferrous salts (1 to 2 g) can cause death in small children.

D. Magnesium

    1. Functions

        a. Binds to the active site of many enzymes

        b. Forms complexes with ATP; MgATP is the species used in most ATP-linked reactions.

    2. Sources: most foods; dairy foods, grains, and nuts (rich sources)

    3. **RDA:** 270 to 320 mg

    4. Deficiency signs and symptoms (see Table 9-6). These are most often seen in alcoholics and patients with fat malabsorption or other malabsorption syndromes.

        a. **Increased neuromuscular excitability,** with muscle spasms and paresthesia; if this is prolonged, tetany, seizures, and coma occur

        b. **Severe hypomagnesemia: depression of PTH release,** which may lead to hypocalcemia

E. Phosphorus (primarily as phosphate)

    1. Functions

        a. 85% of the **phosphorus** in the human body is in the **bone minerals,** calcium phosphate, and hydroxyapatite.

        b. **Phosphates** serve as **blood buffers.**

        c. **Phosphate esters** are **constituents of RNA and DNA.**

        d. **Phospholipids** are the **major constituents of cell membranes.**

2. Sources: seafood, nuts, grains, legumes, and cheeses

3. RDA: 1000 mg

4. Deficiency, which is usually the consequence of abnormal kidney function with reduced reabsorption of phosphate, is **very rare.** Signs and symptoms include (see Table 9-6):

   a. Defective bone mineralization with retarded growth, skeletal deformities, and bone pain.

   b. Diminished release of O_2 from hemoglobin with tissue hypoxia, due to decreased red blood cell 2,3-bisphosphoglycerate.

F. Zinc

   1. Function: essential for the activity of over 200 metalloenzymes

   2. Sources: meat, eggs, seafood, and whole grains

   3. RDA: 12 mg

   4. Deficiency signs and symptoms

      a. Growth retardation and hypogonadism
      b. Impaired taste and smell, poor appetite
      c. Reduced immune function
      d. Mental lethargy
      e. Dry, scaly skin

   5. Zinc toxicity

      a. Ingestion of acidic food or drink from galvanized containers can lead to vomiting and diarrhea.
      b. Inhaling zinc oxide fumes can lead to neurologic damage (**metal fume fever, zinc shakes**).

10

Gene Expression

I. GENETIC INFORMATION

A. **Both DNA and RNA are polynucleotides. Nucleotides,** the monomer units, are composed of three subunits: a **nitrogenous base,** a **sugar,** and **phosphoric acid.**

B. **DNA contains genetic information.** The **genetic code** describes the relationship between the polynucleotide alphabet of **four bases** and the **20 amino acids.** The base sequences in one strand of parental DNA dictate the amino acid sequences of proteins.

C. **Protein synthesis** is an expression of genetic information. The making of proteins involves two processes:

    1. Transcription (DNA to RNA)

    2. Translation (RNA to protein) [Figure 10-1]

D. **The genetic code is universal.** The rules that relate the nucleotide sequence in messenger RNA to the amino acid sequence in proteins (the genetic code per se) are the same in all organisms, with minor exceptions in mitochondria.

    1. A **sense codon, a three-nucleotide codon, specifies each amino acid** (e.g., UUU = phenylalanine, UCU = serine).

    2. Other properties of the genetic code:

        a. It is **contiguous** (i.e., codons do not overlap, and they are not separated by spacers).
        b. It is **degenerate** (i.e., there is more than one codon for some amino acids).
        c. It is **unambiguous.** Each codon specifies only **one** amino acid.

Figure 10-1. Diagram showing the flow of genetic information. *mRNA* = messenger RNA; *tRNA* = transfer RNA; *rRNA* = ribosomal RNA.

E. **Additional information.** Some nucleotides contain genetic information in addition to the sequences that code for polypeptide synthesis.

1. **DNA** contains transcription promoters, binding sites for regulatory proteins, and signals for gene rearrangements.

2. **Messenger RNA (mRNA)** contains transcription terminators, processing signals, translation alignment signals, as well as start and stop signals.

F. **Location** of DNA and protein synthesis

1. In **eukaryotic cells:** replication and transcription occur in the nucleus, translation occurs in the cytosol.

2. In **the human body:** all organs and tissues, except red blood cells.

II. DNA AND RNA: NUCLEIC ACID STRUCTURE

A. **DNA** is a polymer of **deoxyribonucleotides** that are linked by **3′-5′ phosphodiester bonds** (Figure 10-2). The precursors of DNA are deoxyribonucleoside triphosphates (dATP, dGTP, dTTP, and dCTP).

1. **Shape.** DNA is a **double-stranded helix,** with strands that are **antiparallel** and **complementary** (Figure 10-3).

a. **Antiparallel** means that one chain runs in a 5′-to-3′ direction, and the other runs in a 3′-to-5′ direction.

b. **Complementary** means that the **base adenine (A)** always pairs with the **base thymine (T),** and the **base guanine (G)** always pairs with the **base cytosine (C).** There are 10 base pairs per turn.

2. **Stabilizing forces.** The DNA double helix is stabilized by **hydrogen bonds** between the bases on complementary strands and **stacking** and **hydrophobic forces** between bases on the same strand. **AT** base pairs have **two hydrogen bonds,** and **GC** base pairs have **three hydrogen bonds.**

B. **RNA,** which is a polymer of **ribonucleotides,** is also linked by **3′-5′ phosphodiester bonds.**

1. RNA contains the **base uracil (U) instead of T,** as well as A, G, and C.

2. **Shape.** Unlike DNA, RNA is a **single-stranded helix.** Single-stranded RNA may form **internal double-stranded regions,** which are sometimes called **hairpin loops.**

3. There are three classes of RNA: mRNA, ribosomal RNA (rRNA), and transfer RNA (tRNA).

C. **Denaturation.** Nucleic acids in double-stranded form (i.e., DNA or sometimes RNA) **unwind** or **denature** when subjected to high temperatures, pH extremes, and certain chemicals (e.g., formamide, urea).

1. Denaturation causes the **hyperchromic effect,** which is an **increase in ultraviolet (UV) absorption (A_{260}).**

2. Denaturation causes a **decrease in viscosity.**

3. A polynucleotide denatures at a certain temperature, known as the **melting temperature (T_m).** GC-rich regions form more stable double helices than AT-rich regions; thus, GC-rich DNA has a higher T_m than AT-rich DNA.

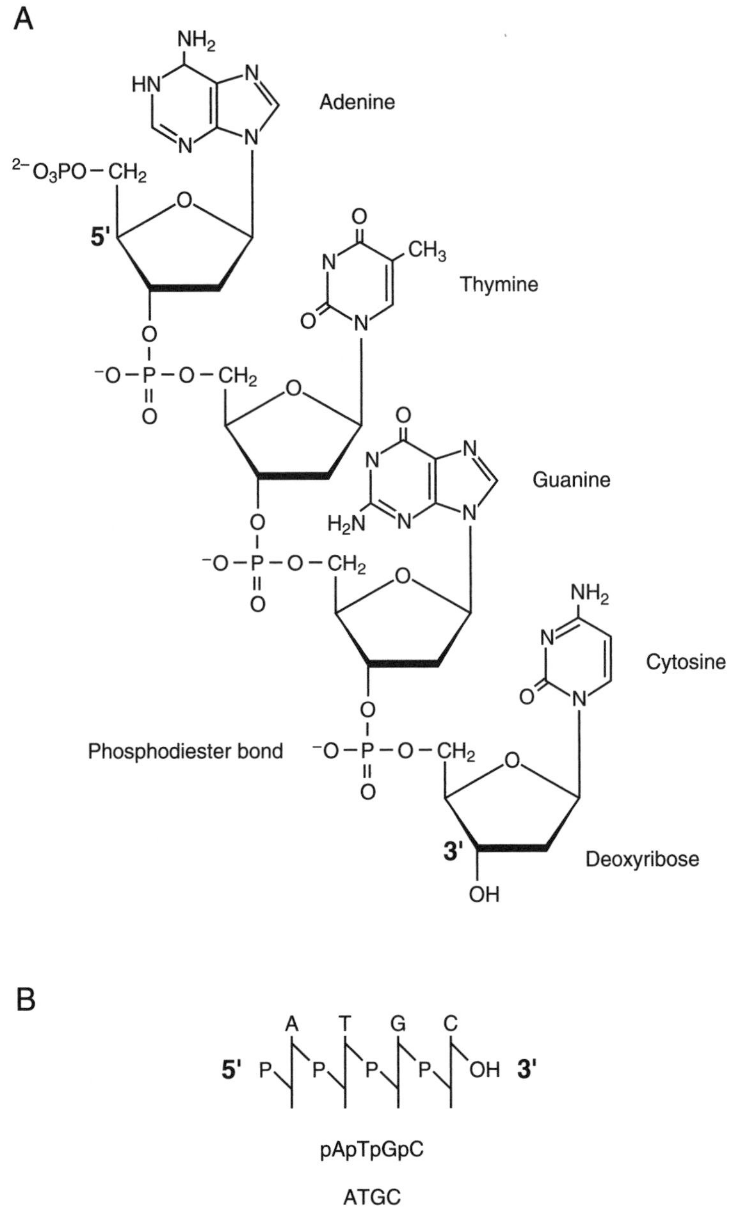

A

B

pApTpGpC

ATGC

Figure 10-2. The structural formula of a deoxyoligonucleotide (A) and abbreviations in DNA (B).

4. When denatured polynucleotides are cooled, or the denaturing agents are removed by dialysis, **complementary** single-stranded regions **reassociate** in a process called **annealing**.

5. Complementary DNA and RNA strands can also associate, or **hybridize**. The presence of DNA or RNA of known sequence may be detected using **hybridization probes**.

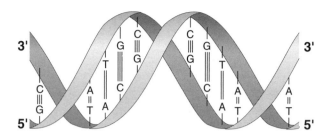

5'–A–T–G–C————3'
3'–T–A–C–G————5'

Figure 10-3. Schematic representation of the DNA double helix showing the two antiparallel, complementary strands.

III. DNA SYNTHESIS (REPLICATION)

A. **Dividing cells** go through an ordered series of events called the **cell cycle.**

1. **Mitosis** is the period when two sets of chromosomes are assembled and **cell division** occurs.

2. Mitosis is followed by **interphase,** which has three subphases: G_1, S, and G_2 (G = gap, S = **synthesis**).

   a. In G_1 **phase,** a cell prepares to initiate DNA synthesis. The chromosomes uncoil and form **euchromatin.**

   b. **DNA synthesis (replication)** occurs during **S phase;** the DNA content doubles. RNA synthesis (transcription) is also at a high level. When DNA synthesis is complete and most other cell constituents have doubled, the cell proceeds into G_2 phase.

   c. During G_2 **phase,** the cell synthesizes the RNA and proteins required for mitosis to occur. Chromatin condenses to form **heterochromatin** and the nuclear membrane disappears.

3. A cell that has completed interphase (G_1, S, G_2) is ready for another round of mitosis.

B. **Replication.** This process is the production of two double-stranded DNA molecules that are identical in every way to the parent DNA.

1. **Replication is semiconservative;** each daughter DNA contains one strand of parental DNA and one newly synthesized daughter strand (Figure 10-4).

2. **Catalysis.** DNA polymerases **(DNAPs)** catalyze DNA synthesis.

   a. **In prokaryotes (e.g., bacteria), DNAP III** is involved in replication.

   b. In eukaryotes, several classes of DNAPs play an important role: α copies the lagging strand, δ copies the leading strand, β carries out repair; and γ carries out mitochondrial DNA replication.

C. **Replication** is a five-step process, each step involving one or more proteins and enzymes.

1. **DNA-unwinding proteins** or **helicases** unwind the DNA duplex. **Topoisomerases** (in *Escherichia coli*, DNA gyrase) relieve the strain imposed by the unwinding.

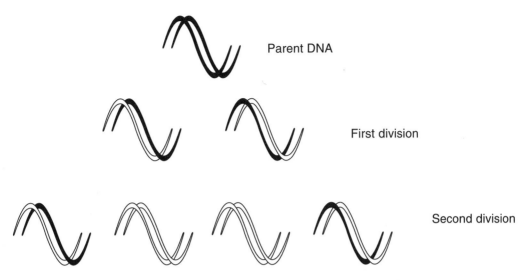

Figure 10-4. Strands of replicated DNA, which demonstrate that DNA replication is semiconservative.

2. Primase makes RNA primers (short pieces of nucleic acid) that are complementary to the DNA template strand. This process moves in the **5′-to-3′ direction** on the newly synthesized primer.

3. At the 3′ end of the primer, **DNAP** adds nucleotides to the 3′-OH, and the so-called **"leading strand" grows continuously in the 5′-to-3′ direction.**

    a. The ability of DNAPs to add incoming nucleotides only to the 3′-OH makes DNA replication **discontinuous** in the other daughter strand, known as the **"lagging strand."**

    b. The segments of newly synthesized DNA, known as **Okazaki fragments,** are then linked to form a continuous DNA chain (Figure 10-5).

4. The DNAP complex removes and replaces the RNA primers. In *E. coli*, DNAP I, which has both **5′-to-3′ exonuclease** and **polymerase** activities, performs this function.

5. DNA ligase joins the ends together.

D. Errors. DNAPs are also associated with **3′-to-5′ exonuclease** activity, which allows detection and removal of mismatched base pairs. This corrective process is called **editing.**

E. DNA repair. UV irradiation, heat, pH extremes, and certain chemicals can alter purines and pyrimidines. A variety of mechanisms are involved in the repair of damaged DNA.

1. Excision repair is a mechanism for correcting thymine dimers in DNA that are created by UV radiation.

    a. A UV-specific endonuclease "nicks" one strand of the double helix, which opens a phosphodiester bond on the 3′ side of the thymine dimer.

    b. DNAP synthesizes a new DNA strand. The **exonuclease activity of DNAP** carries out a 5′-3′ excision of the damaged strand.

    c. DNA ligase joins the ends together.

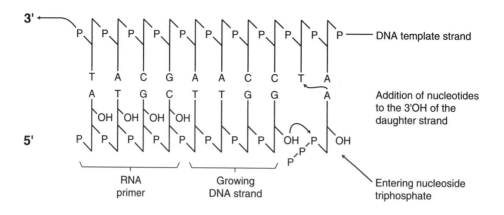

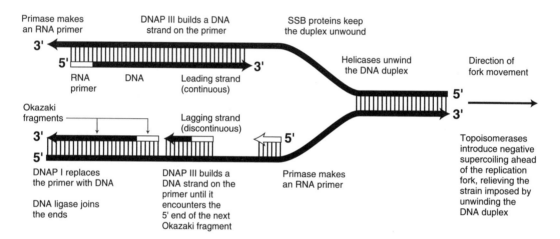

Figure 10-5. *Top*: schematic representation of the action of DNA polymerase. *Bottom*: schematic representation of continuous and discontinuous DNA replication with Okazaki fragments. *DNAP* = DNA polymerase; *SSB* = single-stranded binding.

    2. Diseases resulting from defective excision repair

        a. Xeroderma pigmentosum, a skin disease, is caused by a defective excision repair due to a mutant UV-specific endonuclease. This condition eventually leads to skin cancer.

        b. Ataxia–telangiectasia, Fanconis syndrome (anemia), and **Bloom syndrome** are other diseases associated with defects in excision repair.

IV. TRANSCRIPTION. This process leads to **synthesis of RNA,** with a sequence that is **complementary** to that of the **DNA template.**

    A. Catalysis. RNA polymerases (RNAPs) catalyze transcription.

        1. In prokaryotes, just one kind of RNAP, a large enzyme with many subunits, synthesizes all classes of RNA. The antibiotic **rifampicin** inhibits bacterial synthesis of RNA.

2. In eukaryotes, there are four classes of RNAP, which are all large enzymes with many subunits.

   a. RNAP I synthesizes rRNA. It is found in the nucleolus and is resistant to inhibition by α-amanitin. This inhibitor, which is derived from the poisonous mushroom *Amanita phalloides*, binds to some eukaryotic RNAPs.

   b. RNAP II synthesizes mRNA. This enzyme, found in the nucleoplasm, is highly sensitive to inhibition by α-amanitin [inhibition constant (K_i) = approximately 10^{-9} to 10^{-8} M].

   c. RNAP III synthesizes tRNA and 5S rRNA. This enzyme, which is also found in the nucleoplasm, is moderately sensitive to α-amanitin (K_i = 10^{-5} to 10^{-4} M).

   d. Mitochondrial RNAP, which is inhibited by rifampicin but not by α-amanitin, transcribes RNA from all mitochondrial genes.

B. Transcription cycle. RNA synthesis occurs in four stages (Figure 10-6).

   1. Binding. RNAP binds to specific **promoter sequences** on the DNA, which orients the RNAP on the **sense strand** in a position to begin transcription. A short stretch of the DNA duplex unwinds to form a **transcription bubble.** Both prokaryotic and eukaryotic promoters have **consensus sequences** "upstream" of the start site (Figure 10-7).

   2. Initiation involves the formation of the first phosphodiester bond. ATP or GTP forms a base pair with the template base on the **antisense** strand at the origin, and then the base of the next nucleoside triphosphate pairs with the next template base and forms a phosphodiester bond with the ATP or GTP, eliminating PP_i. Rifampicin blocks initiation in prokaryotes.

   3. Elongation proceeds along the DNA sense strand, with the RNA growing in the 5'-to-3' direction. The DNA duplex reforms behind the enzyme, and the 5' end of the RNA is released as a single strand. **Actinomycin D,** which intercalates between GC sequences in DNA, blocks elongation.

   4. Termination. In prokaryotes, this occurs at the site of a stem-loop (hairpin loop) followed by a string of Us (Figure 10-8). The presence of a *rho* protein makes this process more efficient. In eukaryotes, termination signals are poorly understood.

C. Processing. After transcription, RNA is usually **processed,** or modified. In all organisms, rRNA and tRNA are usually shortened after transcription.

   1. In prokaryotes, mRNA is used as unaltered **primary transcript** as soon as it is made.

   2. In eukaryotes, mRNA is extensively processed. Unprocessed eukaryotic mRNA is sometimes referred to as **heterogeneous nuclear RNA.**

   3. The **"cap"** at the 5' end of mRNA in eukaryotes protects against nuclease digestion and helps align the mRNA properly during translation. It contains **7-methylguanine** in a 5'-5' triphosphate linkage to the 5' ribose (Figure 10-9).

   4. The **poly(A) tail** in eukaryotic mRNA is a string of adenylate residues added to the 3' end (see Figure 10-9).

   5. Many eukaryotic mRNA primary transcripts contain untranslated regions called **intervening sequences,** or **introns** (see Figure 10-9). Removal of introns involves **RNA splicing.**

      a. Introns begin with GU and end with AG. Small nuclear RNAs form base pairs with these splice junctions and assist the splicing enzymes in making a precise cut.

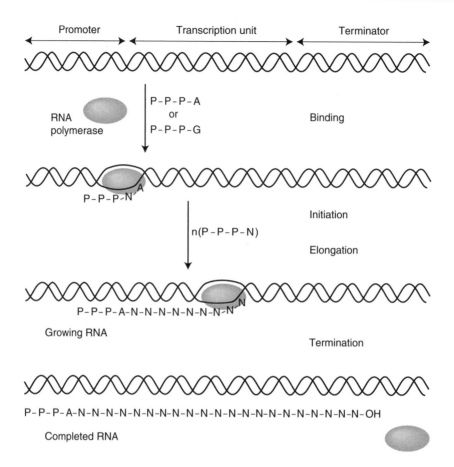

Figure 10-6. Schematic representation of transcription. *P-P-P-A* = ATP; *P-P-P-G* = GTP; *P-P-P-N* = any nucleoside triphosphate.

     b. "Lariat" structures form in the intron, and this intermediate is removed and discarded. Splicing must be very accurate if mRNAs are to be correctly translated into protein.

V. TRANSLATION (PROTEIN SYNTHESIS). This involves the polymerization of amino acids in a precise sequence directed by the sequence of bases in mRNA.

    A. Amino acid activation (initial step). The enzyme **aminoacyl-tRNA synthetase** links amino acids to their specific (cognate) **tRNAs** to form **aminoacyl-tRNAs (AA-tRNAs)** [Figure 10-10].

       1. The **tRNA** is the **adaptor molecule** that brings the base triplet code of nucleic acids together with the amino acid code of proteins.

       2. Each **AA-tRNA synthetase** joins a specific amino acid to the 3′-terminal OH of a specific tRNA.

       3. The tRNA for an amino acid contains a three-base **anticodon** that is **antiparallel and complementary** to the three-nucleotide **codon** in mRNA for that amino acid (e.g., 3′-AAA-5′ is the anticodon for 5′-UUU-3′, the codon for phenylalanine).

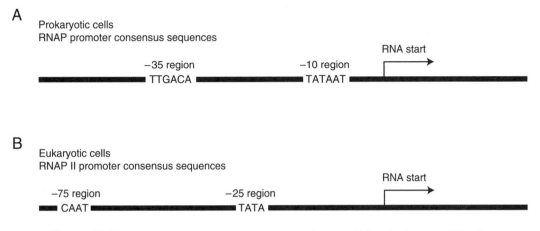

A

Prokaryotic cells
RNAP promoter consensus sequences

RNA start

−35 region   −10 region
TTGACA   TATAAT

B

Eukaryotic cells
RNAP II promoter consensus sequences

RNA start

−75 region   −25 region
CAAT   TATA

Figure 10-7. Promoter consensus sequences in prokaryotic (A) and eukaryotic (B) cells.

A termination sequence in the DNA of a gene:

5'−CCCAGCCCGCCTAATGAGCGGGCTTTTTTTTGA−**3'**
3'−GGGTCGGGCGGATTACTCGCCCGAAAAAAAACT−**5'**

The RNA transcript of the terminator

5'−CCCAGCCCGCCUAAUGAGCGGGCUUUUUUUU−**OH**

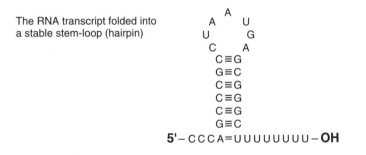

The RNA transcript folded into
a stable stem-loop (hairpin)

```
              A
        A         U
        U         G
         C       A
          C ≡ G
          G ≡ C
          C ≡ G
          C ≡ G
          C ≡ G
          G ≡ C
  5'−CCCA=UUUUUUUU−OH
```

Figure 10-8. A typical prokaryotic termination sequence (terminator).

4. A **high-energy bond** links the amino acid to its tRNA.

B. Site of mRNA-directed protein synthesis: **ribosomes,** which are ribonucleoprotein particles

  1. Composition. One large and one small subunit make up each ribosome. In eukaryotes, the large subunit, which binds AA-tRNA, is 60S, and the small subunit, which binds mRNA, is 40S. Prokaryotic ribosomes are similar, but smaller.

  2. 40S initiation complex. An mRNA, a small ribosomal subunit (40S), eukaryotic initiation factor (eIF) [initiation factor (IF) in prokaryotes], GTP, and methionyl-

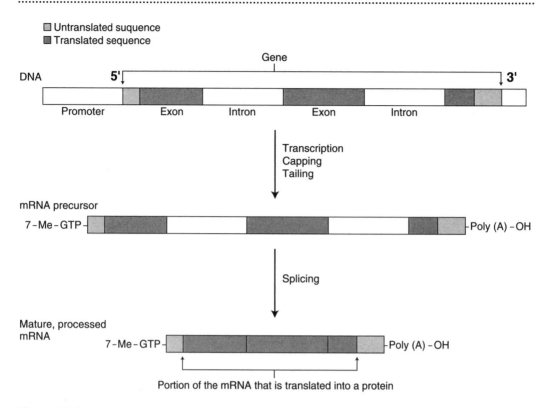

Figure 10-9. Processing in eukaryotic messenger RNA (mRNA) transcription. *7-Me-GTP* = 7-methylguanine in a 5'-5' triphosphate linkage.

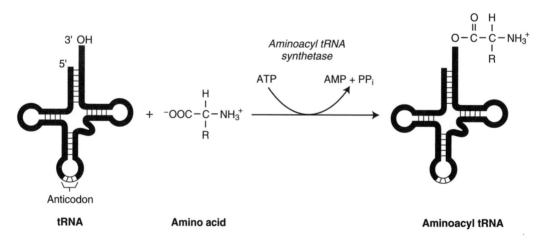

Figure 10-10. Activation of an amino acid to yield an aminoacyl transfer RNA (tRNA), which occurs at the beginning of the translation process.

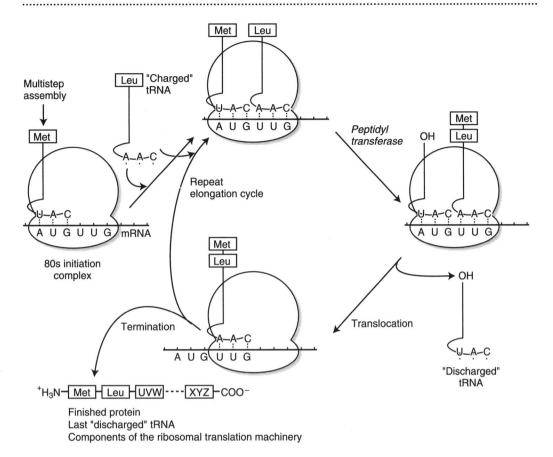

Figure 10-11.   The translation cycle. UVW and XYZ = amino acids.

tRNA form this complex (Figure 10-11). In this process, the hydrolysis of ATP to ADP and P_i occurs.

3. 80S initiation complex. The large ribosomal subunit binds to the 40S initiation complex to form this complex. Methionyl tRNA is positioned at the peptidyl (P) site of the large subunit (80S) [see Figure 10-11], and GDP, P_i, and eIFs are released.

C. **Protein synthesis**

1. Translation (initiation). In eukaryotes, this process begins at the first **AUG** "downstream" (on the 3′ side) of the mRNA cap. AUG, the **translation start** codon, specifies methionine in eukaryotes and N-formylmethionine in prokaryotes (see Figure 10-11). AUG also fixes the **reading frame,** which is the phase in which the sets of three nucleotides are read to produce a protein.

2. Elongation. This occurs in a three-step cycle that repeats each time an amino acid is added (see Figure 10-11).

   a. The **incoming AA-tRNA** binds to the **aminoacyl (A) site** of the large (80S) ribosomal subunit. This requires several protein elongation factors (EFs) and the hydrolysis of GTP.

b. Peptidyl transferase catalyzes the transfer of the amino acid or peptide from the P site to the AA-tRNA on the A site, **with the formation of a peptide bond.** The "uncharged" tRNA dissociates from the complex.

c. The new **peptidyl-tRNA** moves to the P site (i.e., the ribosome moves three nucleotides over on the mRNA), which requires EF-2 and GTP hydrolysis. The **ribosome moves** along the mRNA in the **5'-to-3'** direction, and the **peptide chain grows from the N-terminus to the C-terminus.**

d. After the ribosome has "moved" out of the way, another ribosome can begin translation at the initiation codon. An mRNA with several attached ribosomes that are carrying out translation is known as a **polyribosome or polysome.**

3. Termination. This process occurs when the ribosome encounters a **nonsense (termination) codon** (i.e., UAA, UAG, UGA), which signals **termination** and release of the polypeptide.

a. A protein-releasing factor together with GTP binds to the site.

b. Peptidyl transferase hydrolyzes the peptidyl-tRNA, with the release of the completed polypeptide. Hydrolysis of GTP to GDP and P_i occurs.

c. The ribosomes, which may dissociate into subunits, can be reused.

4. Wobble. The **codon in mRNA (3' base)** and **the anticodon in tRNA (5' base)** [wobble base] can "wobble" at the nucleotide–nucleotide pairing site.

a. In the tRNA anticodon, the wobble base is often inosine, which can pair with U, C, or A in the mRNA codon.

b. In mRNA, G in the wobble position can pair with U or C. U in the wobble position can pair with A or G.

c. Because of wobble, fewer than 61 tRNAs are needed to translate the 61 sense codons of the genetic code.

VI. MUTATIONS

A. **Two principal kinds** of mutation:

1. Substitution of nucleotide for another. The two classes of substitution mutations are (Figure 10-12):

a. Transitions replace a **purine** with a **purine** or a **pyrimidine** with a **pyrimidine.**

b. Transversions replace a **purine** with a **pyrimidine** or vice versa.

2. Insertion or deletion of a nucleotide. These are called **frameshift mutations** when they alter the reading frame.

B. **Missense mutations** specify a **different amino acid; nonsense mutations** convert a normal codon to a terminating codon.

C. The structure of the genetic code tends to minimize the effects of mutation.

1. Changes in the third codon base often do not change the specified amino acid. These are **silent** mutations.

2. Changes in the first codon base generally lead to insertion of the same or a similar amino acid.

3. Amino acids with a strongly nonpolar side chain have codons with pyrimidines in the second position, which means that transition mutations in this base substitute amino acids with similar properties.

4. Amino acids with strongly polar side chains have codons with purines in the second position, so that transitions in this base lead to substitution of amino acids with similar properties.

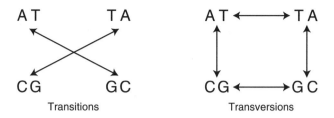

Reading frames in a normal mRNA:

A G C A U G G C U U C U G C G C A G A U U A G G C A C ...

Reading frames in a frameshift mutant mRNA:

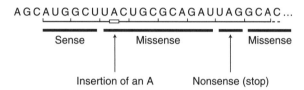

Figure 10-12. Types of mutations.

5. As a general rule, only purine-for-pyrimidine or pyrimidine-for-purine substitutions in the second base of a codon lead to major changes in amino acid side chains.

D. Clinical relevance: osteogenesis imperfecta (OI), a family of diseases characterized by **genetically defective collagen,** leads to abnormal bone fragility. Infants are born with multiple bone fractures.

  1. Mature collagen contains three polypeptide chains that form a left-handed **triple helix.**

  2. The chains are made up of tripeptide repeats of gly-X-Y. **Glycine** at every third position is necessary for proper formation of the triple helix.

  3. OI is frequently the result of a point mutation (i.e., substitution of another amino acid for glycine). For example, substitution of an alanine for glycine at a position near the C-terminal end prevents the formation of the triple helix; in this case, it is a **lethal mutation.**

E. Clinical relevance: sickle cell disease is the consequence of a mutation that substitutes a valine for a glutamic acid at position 6 of the β-chain of hemoglobin A (see Chapter 2 VI A).

F. Clinical relevance: RNA tumor viruses are members of a class of viruses called **retroviruses.**

  1. These viruses have RNA as their genetic material, and contain the enzyme **reverse transcriptase.**

  2. After the virus enters the host cell, reverse transcriptase makes a **DNA copy of its RNA,** forming a DNA–RNA hybrid.

3. Then the reverse transcriptase uses the DNA–RNA hybrid to make a DNA double-helix copy of its own RNA.

4. The virus also contains an **integrase** enzyme that inserts the viral DNA into the host-cell chromosome.

5. Some of the viral genes are **oncogenes,** modifications of the normal host-cell genes that transform the host cells into cancer cells.

6. Human immunodeficiency virus (HIV), which causes **acquired immunodeficiency syndrome (AIDS),** is a retrovirus.

    a. In the case of HIV, the virus infects the **helper T cells** of the immune system, and eventually kills them. This renders the infected person highly susceptible to infections.

    b. The HIV provirus can exist in a latent state within infected cells for a long time, until some (unknown) event activates it. This makes an HIV infection very difficult to treat with drugs.

11

Biotechnology

I. PROTEIN PURIFICATION. Proteins must be purified before they can be studied.

   A. Proteins occur in complex mixtures, which contain many different proteins.

   B. Several **different separation methods** are used to yield a purified protein sample, including:

   1. **Selective precipitation,** which involves use of pH, heat, or salts (e.g., ammonium sulfate) to separate proteins from solutions

   2. **Gel filtration** and **polyacrylamide gel electrophoresis,** which separate proteins on the basis of size

   3. **Gel electrophoresis** and **ion-exchange chromatography,** which separate proteins on the basis of ionic charge

   4. **Affinity chromatography,** which involves removal of proteins from a mixture by specific binding to their ligands or to antibodies

II. PROTEIN ANALYSIS

   A. **Ion-exchange chromatography** is used to determine the **amino acid composition** of a protein after its hydrolysis in HCl.

   B. The **Edman degradation method** is used to determine the **amino acid sequence** (primary structure).

   1. This technique involves the use of phenyl isothiocyanate to label and then remove amino acids one at a time from the N-terminal end (Figure 11-1). This portion of the procedure has been automated.

   2. The Edman degradation method can sequence peptides of only about 50 residues or less, so **proteins must be broken into defined fragments of manageable size.** Some chemicals or enzymes selectively cleave proteins.

   a. **Cyanogen bromide** cleaves peptide bonds on the carboxyl side of methionine residues.

   b. **2-Nitro-5-thiocyanobenzene** cleaves peptide bonds on the amino side of cysteine residues.

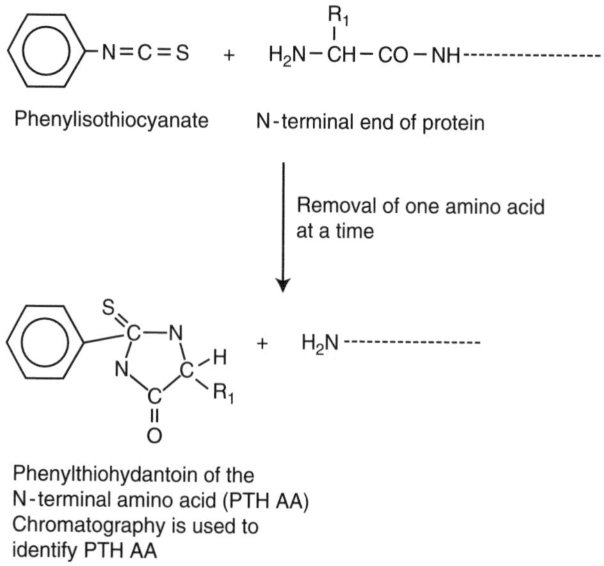

Figure 11-1. Schematic representation of the Edman degradation method for determining the amino acid sequence of a protein. After one amino acid has been removed, the cycle can be repeated about 50 times.

    c. The pancreatic enzyme **trypsin** cleaves peptide bonds on the carboxyl side of lysine or arginine residues.

    d. The enzyme **chymotrypsin** cleaves peptide bonds on the carboxyl side of aromatic and some other bulky nonpolar residues.

  C. **X-ray crystallography** and **nuclear magnetic resonance** are used to determine the **three-dimensional structure.**

III. DNA ANALYSIS

  A. The DNA of a human chromosome contains about 10^8 base pairs, and before this DNA can be studied, fragmentation must occur. **Fragmentation is used to break DNA into reproducible pieces of manageable size.**

    1. Bacterial enzymes known as **restriction endonucleases** are used to cleave DNA at specific palindromic **restriction sites** of four to eight base pairs (Figure 11-2).

    2. Each restriction endonuclease cleaves a DNA molecule into a limited number of fragments of specific and reproducible sizes.

  B. **Gel electrophoresis** is used to separate the DNA fragments on **the basis of size.**

  C. **Southern blotting** is used to detect DNA fragments that contain a **specific base sequence.** Procedural steps include (Figure 11-3):

    1. Denaturing the DNA in the gel with alkali or heat

    2. Transferring ("blotting") the DNA fragments from the gel to a nitrocellulose filter in a way that preserves the pattern of fragments in the gel

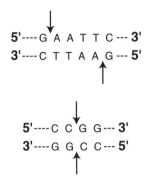

Figure 11-2.   Palindromic restriction endonuclease cleavage sites. In each pair, the lower strands have the same sequence (reading) from 5′ to 3′ as the upper strands. The arrows point to the phosphodiester bonds that are cleaved. *Top:* the site for *Eco*R1, which has overlapping, sticky ends. *Bottom:* the site for *Hae*III, with nonoverlapping, blunt ends.

3. Immersing the nitrocellulose filter in a solution that contains hybridization probes. These probes are oligodeoxynucleotides that complement the specific DNA sequence of interest and have been labeled (e.g., radioactive or fluorescent group).

4. Washing the filter to remove excess probe after allowing sufficient time for the probe to hybridize (anneal) to the complementary DNA

5. Visualizing the spots containing the DNA of interest using autoradiography or fluorescence

D. Northern blotting uses hybridization probes to detect **RNA fragments.** Otherwise, this procedure is performed like Southern blotting.

E. Western blotting uses antibodies to detect **proteins.**

F. The **Sanger dideoxynucleotide method** (chain-termination) and the specific chemical cleavage procedure (Maxam-Gilbert) are two techniques used to **determine the sequence of bases in DNA fragments.** The Sanger dideoxynucleotide method, which is used more frequently, involves the following steps (Figure 11-4):

1. Separating (denaturing) the DNA fragment into single strands and dividing them into four samples

2. Adding the following to each sample: an oligonucleotide **primer,** a large excess of all four **deoxynucleoside triphosphates** (dNTPs: dATP, dGTP, dCTP, dTTP), **DNA polymerase,** and a small amount of a different **dideoxynucleoside triphosphate (ddNTP)** analogous to one of the four DNA nucleotides

   a. To enable detection of the DNA fragments, label the primer at the 5′ end or include a labeled dNTP in the reaction mixture.

   b. The ddNTP stops transcription when it is incorporated into the growing chain, because it has no 3′OH.

3. Subjecting the reaction products to gel electrophoresis and autoradiography, and reading the sequence from the band patterns

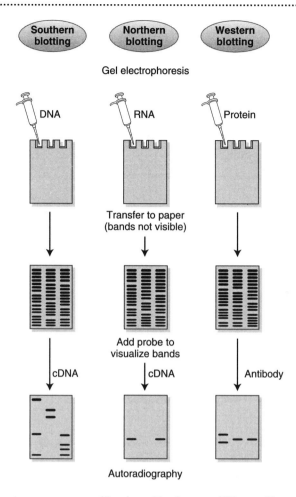

Figure 11-3.   Schematic representation of Southern, Northern, and Western blotting. *cDNA* = complementary DNA. (Adapted with permission from Marks D, Marks A: *Basic Medical Biochemistry*. Baltimore, Williams & Wilkins, 1996, p 244.)

    G. The **polymerase chain reaction (PCR)** is used to amplify very tiny (1–10 ng) pieces of DNA (Figure 11-5).

        1. **Required components**

            a. **DNA to be amplified**

            b. Two oligonucleotide **primers** complementary to the base sequences on each strand of the DNA, one on either end of (flanking) the region to be amplified

            c. All four **dNTPs**

            d. A **heat-stable DNA polymerase** (usually *Taq* polymerase)

        2. The **PCR process** involves the following steps:

            a. Mixing the DNA, a large excess of the primers, the dNTPs, and the polymerase

            b. Heating the mixture briefly to 90°C to separate the DNA strands

            c. Cooling the mixture to 60°C, allowing the primers to anneal to the DNA and the polymerase to extend the chains

            d. Repeating steps b and c 20 to 30 times

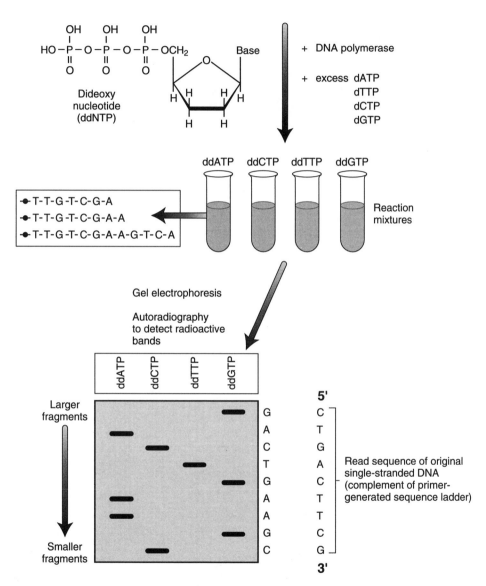

Figure 11-4. Schematic representation of the sequencing of a DNA fragment by the Sanger dideoxynucleotide method. *ddNTP* = dideoxynucleoside triphosphate. (Adapted with permission from Gelehrter TD, Collins FS: *Principles of Medical Genetics.* Baltimore, Williams & Wilkins, 1995.)

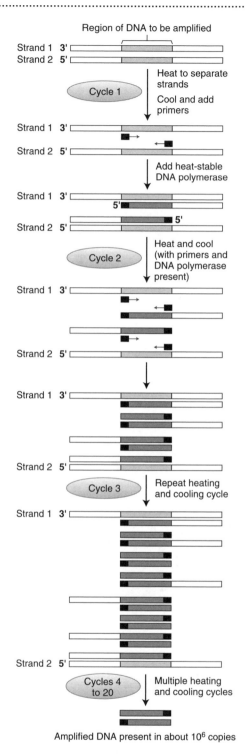

Figure 11-5. Schematic representation of the polymerase chain reaction (PCR). (Adapted with permission from Marks D, Marks A: *Basic Medical Biochemistry*. Baltimore, Williams & Wilkins, 1996, p 248.)

H. DNA fingerprinting can be used to identify an individual's DNA and to trace a family tree. DNA from each individual has a characteristic DNA fingerprint.

   1. DNAs from different individuals contain sequence variations known as **polymorphisms,** which may involve an insertion or a deletion of one or more bases or a change in the sequence of bases.

   2. Some sequence polymorphisms occur in or near the sites of cutting by restriction enzymes. This leads to **restriction fragment length polymorphisms (RFLPs),** which are differences in the sizes of restriction fragments between individuals.

   3. Southern blotting can be used to visualize RFLPs.

IV. CLONING OF RECOMBINANT DNA AND PROTEIN. Cloning involves insertion of DNA fragments into a vector (e.g., bacteriophage, plasmid) that will replicate within a bacteria (Figure 11-6).

   A. Cloning may be used to amplify a DNA sample and obtain a large quantity for further study (e.g., sequencing).

   B. Under the proper conditions, recombinant DNA inserted into a target cell population can be transcribed and translated, which means it is expressed as the protein gene product.

   C. Cloning involves the following steps:

   1. Cleaving the DNA that is to be cloned and that of the vector with a restriction endonuclease

   2. Attaching the foreign DNA to the vector by treatment with DNA ligase, thus producing **chimeric** or **recombinant** DNA

   3. Allowing bacterial cells to be transformed with the vector containing the recombinant DNA and then plating them to produce individual colonies

   4. Identifying and selecting the colonies containing the cloned recombinant DNA using a probe (e.g., DNA, RNA, antibody) and then isolating and culturing them

   5. Isolating and characterizing the cloned DNA or the protein expressed from the cloned DNA from the bacterial cells

   D. Bacteria are not suitable for expressing mammalian proteins because they lack the enzymes for processing eukaryotic mRNAs. Two approaches to generating recombinant mammalian proteins are:

   1. Copying the mRNA for the protein with reverse transcriptase to produce a complementary DNA. This process involves:

   a. Ligating the cDNA into a bacterial expression vector
   b. Transforming a bacterial strain
   c. Allowing the transformed bacteria to express the protein
   d. Isolating and purifying the protein

   2. Ligating the mammalian DNA into a eukaryotic expression vector, which involves:

   a. Transforming yeast or cultured mammalian cells
   b. Allowing the cells to express the protein
   c. Isolating and purifying the protein

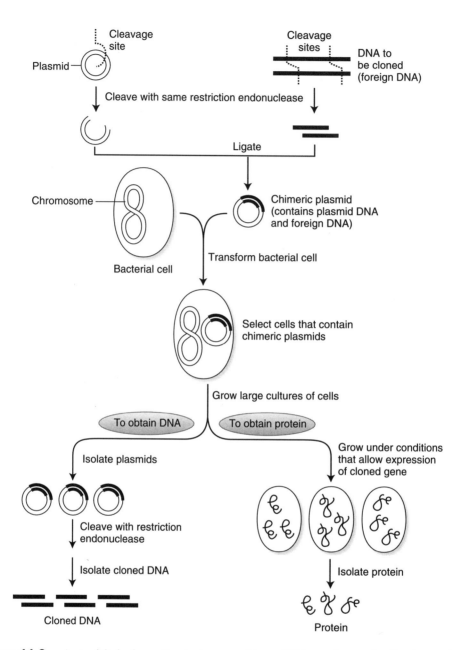

Figure 11-6. A simplified scheme for cloning recombinant DNA in a bacterial cell culture. The cloned DNA can be replicated to obtain a large quantity of DNA for study, or expressed to obtain a large quantity of the gene product, usually a protein. (Adapted with permission from Marks D, Marks A: *Basic Medical Biochemistry*. Baltimore, Williams & Wilkins, 1996, p 247.)

V. CLINICAL RELEVANCE

A. Diabetic syndromes

1. Diabetes insipidus is a condition resulting from the failure of the posterior pituitary to secrete sufficient **antidiuretic hormone** (arginine **vasopressin, AVP**), a polypeptide. Polyuria (excreting large volumes of urine) and polydipsia (drinking large volumes of water) are characteristic. Treatment includes administration of the **synthetic AVP analogue desmopressin** (1-deamino-8-D-arginine vasopressin).

2. Diabetes mellitus is a disorder that leads to abnormalities of carbohydrate (e.g., hyperglycemia) and fat metabolism. Treatment involves daily injections of the protein **insulin.** The supply of beef or pork insulin is necessarily limited, and **recombinant human insulin** is now available in unlimited amounts.

B. **Pituitary dwarfism** requires treatment with **human growth hormone (HGH).** Recombinant **HGH** has replaced HGH extracted from human cadavers.

C. **Hematologic problems. Recombinant tissue plasminogen activator,** an enzyme, is useful for **dissolving blood clots** (e.g., in coronary arteries after a heart attack).

12
Hormones

I. OVERVIEW

A. The **endocrine system** consists of a group of **endocrine glands** that secrete **hormones** into the bloodstream.

B. These **hormones,** which travel to all parts of the body, exert specific catalytic effects on **target tissues** (Figure 12-1).

II. CLASSIFICATION OF HORMONES

A. **Water-soluble hormones** (Figure 12-2)

1. **Catecholamine hormones** (e.g., epinephrine)

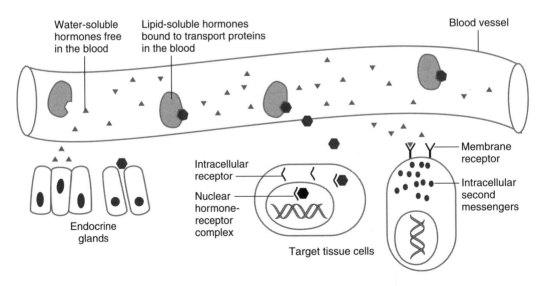

Figure 12-1. Diagrammatic representation of the relationship between the endocrine glands and their target tissues.

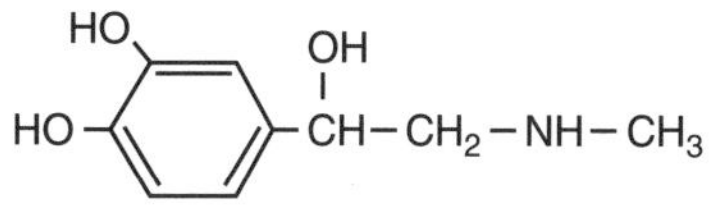

Epinephrine

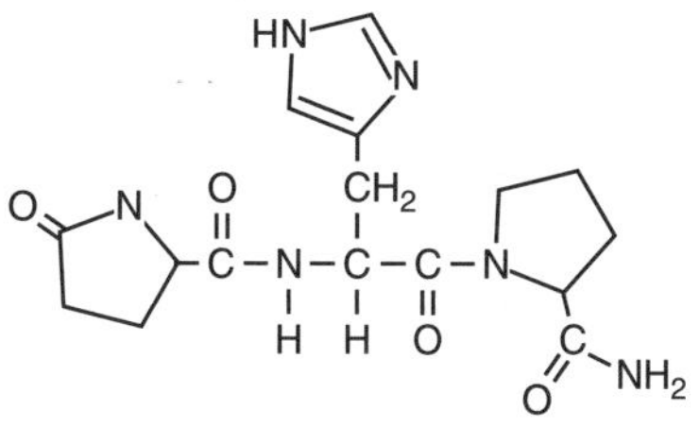

Thyrotropin-releasing hormone (TRH)

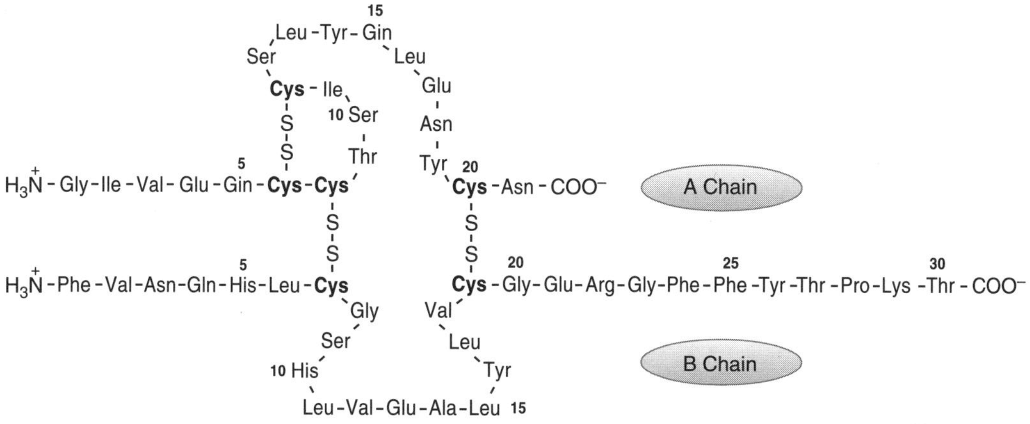

Human insulin

Figure 12-2.   Structure of selected water-soluble hormones. (Adapted with permission from Marks D, Marks A: *Basic Medical Biochemistry*. Baltimore, Williams & Wilkins, 1996, p 96.)

    2. Peptide hormones (e.g., TRH)

    3. Protein hormones (e.g., insulin)

  B. Lipid-soluble hormones (Figure 12-3)

    1. Steroid hormones (e.g., cortisol, testosterone, estradiol)

    2. Thyroid hormones (e.g., thyroxine)

III. MECHANISMS OF HORMONE ACTION

  A. Water-soluble hormones

    1. These hormones bind to **membrane receptors** in their target tissues. Receptor binding leads to the production of **intracellular second messengers** (see Figure 12-1).

    2. Hormone binding to one group of receptors stimulates **adenylate cyclase**, which converts ATP to **adenosine 3′,5′-monophosphate (cAMP).** This cAMP activates **protein kinase A,** an enzyme that phosphorylates several proteins.

      a. Some of the proteins are **enzymes,** and phosphorylation may have either positive or negative effects on their activity.

      b. Some of the proteins are transcription factors called **cAMP-responsive element-binding proteins,** which alter gene expression.

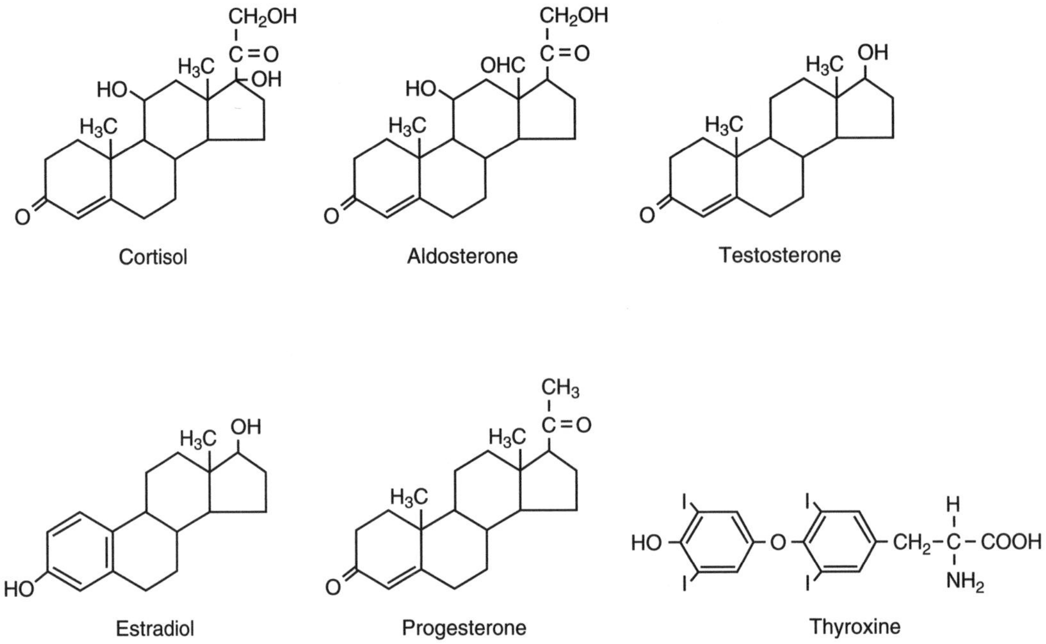

Figure 12-3. Structure of selected lipid-soluble hormones.

3. Hormone binding to a second group of receptors activates **phospholipase C,** which hydrolyzes **phosphatidylinositol 4,5-bisphosphate (PIP₂)** to yield **inositol 1,4,5-trisphosphate (IP₃) and diacylglycerol (DAG).**

   a. DAG activates **protein kinase A.**

   b. IP₃ stimulates the release of Ca^{2+} from the endoplasmic reticulum into the cytosol, where it modulates several enzyme activities.

4. Hormone binding to a third group of receptors stimulates **tyrosine kinase** activity, which involves **autophosphorylation** of some of the receptor's own tyrosine residues. The phosphorylated receptors then interact with other intracellular proteins to alter cell activities.

B. **Lipid-soluble hormones**

   1. These hormones pass through the cell membrane and bind to **intracellular hormone receptor proteins** (see Figure 12-1).

   2. Hormone binding **activates** the receptors, converting them to **transcription factors,** which bind to **hormone response elements** in DNA and **alter gene expression.**

IV. HORMONES THAT REGULATE FUEL METABOLISM

A. **Insulin,** which is secreted by the β-cells in the pancreatic islets of Langerhans, is a small protein with two polypeptide chains connected by disulfide bonds (see Figure 12-2).

   1. **Actions.** Insulin acts on **adipose tissue, skeletal muscle,** and **liver** to **lower blood glucose** and **nonesterified fatty acid** concentrations. It leads to:

   a. **Increased glucose entry** into adipose tissue and muscle

   b. **Increased glucose metabolism** in adipose tissue, muscle, and liver

c. **Increased amino acid entry** into muscle
d. **Decreased lipolysis** and **fatty acid release** in adipose tissue

2. **Secretion. High levels of blood glucose** (hyperglycemia) and **amino acids increase** insulin secretion, whereas **epinephrine decreases** it.

B. **Glucagon,** which is secreted by the α-cells of the pancreatic islets, is a protein.

1. **Actions.** Glucagon **increases blood glucose and fatty acid concentration** by stimulating production of cAMP and activation of protein kinase A in **liver, adipose tissue, and muscle.** This leads to:
   a. **Increased glycogenolysis** in liver and muscle
   b. **Increased lipolysis** and **fatty acid release** in adipose tissue
   c. **Increased functioning of the glucose–alanine cycle** between liver and muscle

2. **Secretion. Low levels of blood glucose** (hypoglycemia) and **high levels of blood amino acids** both **increase glucagon secretion.**

C. **Epinephrine,** which is secreted by the **adrenal medulla,** is a **catecholamine** (see Figure 12-2).

1. **Actions.** Epinephrine **elevates blood glucose** and **fatty acids** and **provokes the fight-flee reflex** by stimulating production of cAMP and activation of protein kinase A in its target tissues. This leads to:
   a. **Increased glycogenolysis** in muscle and liver
   b. **Increased lipolysis** and **fatty acid release** in adipose tissue
   c. **Development of the fight-flee reflex,** with increased heart rate (chronotropic effect) and force of contraction (inotropic effect), dilation of blood vessels in skeletal muscle, and constriction of blood vessels in the skin and splanchnic bed

2. **Secretion. Hypoglycemia, low oxygen tension** (hypoxia), **and neural factors stimulate** epinephrine secretion.

D. **Cortisol,** which is secreted by the adrenal cortex, is a **steroid** (glucocorticoid) hormone (see Figure 12-3).

1. **Actions.** Cortisol has many functions, some of which lead to **elevation of blood glucose.**
   a. **Increased muscle protein breakdown,** which releases amino acids as substrates for gluconeogenesis
   b. **Increased synthesis of gluconeogenic enzymes** in the liver
   c. **Inhibition of insulin action**
   d. **Increased total body fat** at the expense of muscle protein
   e. **Increased water excretion** by the kidney
   f. **Inhibition of inflammation**
   g. **Suppression of the immune system**
   h. **Increased resistance to stress**

2. **Secretion. Adrenocorticotrophic hormone (ACTH),** a pituitary hormone, regulates cortisol secretion. **Corticotropin-releasing hormone (CRH)** from the hypothalamus leads to ACTH secretion. Reduced CRH secretion results from elevated plasma cortisol.

V. HORMONES THAT REGULATE SALT AND WATER BALANCE

A. Aldosterone, which is secreted by the adrenal cortex, is a steroid hormone.

   1. Actions. Aldosterone stimulates Na$^+$ retention and K$^+$ secretion by the kidney, sweat glands, and intestinal mucosa.

   2. Secretion. The renin–angiotensin system and elevated blood K$^+$ both stimulate aldosterone secretion.

B. Arginine vasopressin (AVP) [also known as antidiuretic hormone], which is secreted by the posterior pituitary, is a small peptide.

   1. Actions. AVP stimulates water reabsorption by the kidney.

   2. Secretion. High plasma osmolality and neural impulses both stimulate AVP secretion.

VI. HORMONES THAT REGULATE CALCIUM AND PHOSPHATE METABOLISM

A. Parathyroid hormone (PTH), which is secreted by the parathyroid glands, is a protein.

   1. Actions. PTH raises plasma calcium and lowers plasma phosphate. Other functions include:

   a. Stimulation of osteoclasts, leading to dissolution of bone salts and release of Ca^{2+} and PO$_4^{3-}$ into the blood

   b. Decreased Ca^{2+} excretion and increased PO$_4^{3-}$ excretion by the kidney

   c. Stimulation of calcitriol formation from 25OH-D$_3$ by the kidney, thus leading to increased calcium absorption from the intestine

   4. Secretion. Hypocalcemia stimulates secretion of PTH, and hypercalcemia inhibits it.

B. Calcitriol, or 1,25-dihydroxycholecalciferol [1,25(OH)$_2$-D$_3$], is derived from vitamin D$_3$.

   1. Synthesis

   a. Vitamin D$_3$ is converted to calcitriol by two hydroxylation reactions, one in the liver and one in the kidney.

   b. Hypocalcemia and PTH stimulate the second hydroxylation reaction.

   c. Calcitriol is released from the kidney into the circulation.

   2. Actions

   a. Stimulation of calcium absorption from the gut

   b. Increased efficiency of PTH action on bone

VII. HORMONES THAT REGULATE BODY SIZE AND METABOLISM

A. Thyroxine and triiodothyronine, which are secreted by thyroid follicle cells, are iodoamino acids (see Figure 12-3).

   1. Actions. Elevated thyroid hormone causes an increase in metabolic rate throughout the body, which is manifested by:

   a. Increased heat production

   b. Increased growth

   c. Increased mental activity

   d. Increased sensitivity to epinephrine

   e. Increased catabolism of cholesterol, leading to decreased blood cholesterol